ez101 study keys

Statistics

Second Edition

Martin Sternstein, Ph.D.
Professor of Mathematics
Ithaca College
Ithaca, New York

BARRON'S

All inquiries should be addressed to:
Barron's Educational Series, Inc.
250 Wireless Boulevard
Hauppauge, New York 11788
http://www.barronseduc.com

Library of Congress Catalog Card No. 2005041080

ISBN-13: 978-0-7641-2915-5
ISBN-10: 0-7641-2915-5

Library of Congress Cataloging-in-Publication Data

Sternstein, Martin.
    Statistics / Martin Sternstein.—2nd ed.
        p. cm.—(Barron's EZ 101 study keys)
    Includes index.
    ISBN 0-7641-2915-5
    1. Statistics. I. Title. II. Series.

QA276.18.S74 2005
519.5—dc22

2005041080

PRINTED IN CANADA

9 8 7 6 5 4 3 2 1

CONTENTS

To the reader

This book is intended as an overview and supplementary learning aid for college students taking an introductory course in statistics or needing a self-review guide of elementary statistics as a prerequisite for another course. This book is also helpful for students preparing to take the AP Statistics exam, although *Barron's How to Prepare for the AP Advanced Placement Exam Statistics* is more comprehensive for this purpose. The student should be able to *self-read* this guide and find it useful when doing homework or studying for exams, and in clearing up basic points when the standard textbook might use extensive theoretical discussion.

The material is divided into nine major **Themes** and then further into seventy-three units called **Keys**, each focusing on a single topic. Objectives are made very explicit, and there are carefully worked-out examples. Special emphasis has been given to word problems showing the usefulness of statistics in a variety of disciplines.

Some of the problems, especially those involving binomial and Poisson probabilities, require a scientific calculator with an x^y-key. A few of the problems involve small data sets, and a "stat" mode on a calculator will shorten calculations. Most helpful is a graphing calculator such as the TI-83, and alternative calculations using the TI-83 statistical package are referred to as appropriate. And of course if one is working on a large data project, access to a computer statistics package is almost a necessity. This review book will help students with the basic concepts, language, and techniques of statistics so as to better understand the computer calculations.

I wish to thank Dora Daniluk of Mayde Creek High School in Houston, Texas for many useful suggestions; my students who tested and commented on the examples; my sons Jonathan and Jeremy, who always reminded their dad of what is most important in life; and my wife, Faith, without whose love and encouragement this project would not have been possible.

Ithaca College   Martin Sternstein
August 2005

Theme 1   DESCRIPTIVE STATISTICS

*T*he need to make sense of masses of information has led to formalized ways of describing the tremendous and ever-growing quantity of numerical data being collected in almost all areas of knowledge. Given a raw set of data, there is often no apparent overall pattern. Perhaps some values are more frequent, sometimes a few extreme values stand out, and usually the range of values is noticeable. Presenting data involves such concepts as representative or average values, measures of dispersion, and positions of various values, all of which fall under the broad topic of *descriptive statistics,* the subject of Theme 1. This is in contrast to *statistical analysis,* which draws inferences from limited data, a subject that will be discussed in later themes.

Key 1   Central tendency

OVERVIEW   *The word* average *arises in phrases common to everyday conversation, from* batting averages *to* average life expectancies, *and has come to mean a "representative score" or a "typical value" or the "center" of a distribution. Mathematically, there are a variety of ways to define the center of a set of data, and in this key we consider the three most common methods.*

The median: Derived from the Latin *medius,* meaning "middle," the **median** is the middle number of a set of numbers arranged in numerical order. (If there is an even number of values, the median is the result of adding the two middle values and dividing by 2.)
- The median of the set $\{2, 3, 6, 6, 7, 9, 10, 13, 25\}$ is 7.
- The median of the set $\{2, 3, 4, 7, 8.4, 9, 35, 46\}$ is $\frac{1}{2}(7 + 8.4) = 7.7$.

     The median is not affected by exactly how large the larger values are or by exactly how small the smaller values are. Thus it is a particularly useful measurement when the extreme values, called *outliers,* are in some way suspicious, or if one wants to diminish their effect. We say that the median is *resistant* to extreme values. For example, if ten mice try to solve a maze, and nine take less than 15 minutes, while one is still trying after 24 hours, then the most representative value is the median. In certain other situations the median offers the most economical and quickest technique to calculate an average. For example, suppose 10,000 lightbulbs of a particular brand are installed in a factory. An average life expectancy for the bulbs can most easily be obtained by noting how much time passes before exactly one-half of them have had to be replaced. The median is also useful in certain kinds of medical research. For example, to compare the relative strengths of different poisons, a scientist notes what dosage of each poison will result in the deaths of exactly one-half the test animals. This median lethal dose is not influenced by one of the animals being especially susceptible to a particular poison.

The mode: The **mode,** or most frequent value, is an easily understood representative score. It is clear what is meant by "the most common family size is 4" or "the professor gives out more B's than any other grade."

- The mode of $\{2, 3, 6, 6, 7, 9, 10, 13, 25\}$ is 6.
  When two scores have equal frequency, and this frequency is higher than any other, we say that the set is *bimodal*.
- The set $\{2, 3, 6, 6, 7, 7, 10, 13, 25\}$ is bimodal, with modes 6 and 7.

The mean: Summing the values or scores, and dividing by the number of values or scores, gives the **mean,** the most important measure of central tendency for statistical *analysis*.
- $\{2, 3, 6, 6, 7, 9, 10, 13, 25\}$ has mean $(2 + 3 + 6 + 6 + 7 + 9 + 10 + 13 + 25)/9 = 81/9 = 9$.
- The mean of a *whole population* (complete set of items of interest) is often denoted by the Greek letter μ (mu), while the mean of a *sample* (a part of a population) is often denoted as $\bar{x}$.
- The mean average value of the set of all houses in the United States might be $\mu = \$56,400$, while the mean average of 100 randomly chosen houses might be $\bar{x} = \$52,100$ or perhaps $\bar{x} = \$63,800$, or even $\bar{x} = \$124,000$.
- If a sum is formed by selecting one element from each of two sets, then the mean of all such sums is simply the sum of the means of the two sets. Notation: $\mu_{X+Y} = \mu_X + \mu_Y$

In statistics one learns how to estimate a population mean from a sample mean. Throughout this book, *sample* implies *simple random sample;* that is, not only does every element of the population have an equal chance of being included, but also the sample is selected in such a way that every possible sample of the desired size has an equal chance of being included. In the real world, this process of *random* selection is often very difficult to achieve.

Notation: $\mu = \bar{x} = (\Sigma x)/n$, where Σx represents the sum of all the elements of the set under consideration, while n is the actual number of elements (Σ is the upper-case Greek letter sigma).

Key 2 Variability

OVERVIEW *In describing a set of numbers, not only is it useful to designate an average score, but also it is important to be able to indicate the **variability** or the **dispersion** of the measurements. A producer of time bombs aims for small variability—it would not be good if his 30-minute fuses actually ranged from 10 minutes to 50 minutes before detonation. On the other hand, a teacher interested in distinguishing the better from the poorer students aims to design exams with large variability in results—it would not be helpful if all the students scored exactly the same. The players on two basketball teams might have the same average height, but this fact doesn't tell the whole story—if the dispersions are quite different, one team might have a 7-foot player, whereas the other might have no one over 6 feet. Two Mediterranean holiday cruises might advertise the same average age for their passengers, but one could have only passengers between 20 and 25 years old, while the other had only middle-aged parents in their 40's together with their children under age 10.*

The range: The difference between the largest and smallest values of a set is called the **range**. Although it is the most easily calculated measure of variability, the range is entirely dependent on two extreme values and is insensitive to what is happening between. One use of the range is to consider samples with very few items. For example, some quality-control techniques involve taking periodic small samples and basing further action upon the range found in several samples.

The variance: This measure of variability indicates dispersion around the mean. The **variance** of a population, denoted by σ^2 (σ is the lowercase Greek letter sigma), is the average of the squared deviations from the mean:

$$\sigma^2 = \frac{\sum(x - \mu)^2}{n}$$

The variance of a sample, denoted as s^2, is calculated as follows:

$$s^2 = \frac{\sum(x - \overline{x})^2}{n - 1}$$

The standard deviation: The square root of the variance is called the **standard deviation,** where σ represents the standard deviation of a population, and s represents the standard deviation of a sample.

- If $X = \{2, 9, 11, 22\}$, then

$$\mu = \frac{2+9+11+22}{4} = 11$$

$$\sigma^2 = \frac{(2-11)^2 + (9-11)^2 + (11-11)^2 + (22-11)^2}{4} = 51.5$$

and $$\sigma = \sqrt{51.5} = 7.176$$

An alternative arithmetical tool for calculating the variance and standard deviation comes from these equations:

$$\sigma^2 = \frac{\sum x^2}{n} - \mu^2 \text{ and } s^2 = \frac{\sum x^2 - \frac{(\sum x)^2}{n}}{n-1}$$

- For the above $\{2, 9, 11, 22\}$ we could also have calculated

$$\sum x^2 = 2^2 + 9^2 + 11^2 + 22^2 = 690$$

so

$$\sigma^2 = \frac{690}{4} - 11^2 = 172.5 - 121 = 51.5$$

The interquartile range (IQR): This is one method of removing the influence of extreme values on the range. It is calculated by arranging the data in numerical order, removing the upper and lower one-quarter of the values, and then noting the range of the remaining values. That is, it is the range of the middle 50% of the values.

Relative variability: A comparison of two variances may be more meaningful if the means of the populations are also taken into consideration. **Relative variability** is defined to be the quotient obtained by dividing the standard deviation by the mean. Usually it is then expressed as a percentage.

Key 3   Effect of changing units

OVERVIEW   *Changing units, for example, from dollars to rubles or from miles to kilometers, is common in a world that seems to become smaller all the time. It is instructive to note how measures of center and spread are affected by such changes. Adding the same constant to every value increases the mean and median by that same constant; however, the distances between the increased values stay the same, so the range and standard deviation are unchanged. Multiplying every value by the same constant multiplies the mean, median, range, and standard deviation all by that constant.*

KEY EXAMPLE

A set of experimental measurements of the boiling point of an unknown liquid yield a mean of 95.42 degrees Celsius with a standard deviation of 2.48 degrees Celsius. If all the measurements are converted to the Kelvin scale, what are the new mean and standard deviation?

Answer: Kelvins are equivalent to degrees Celsius plus 273.16. The new mean is thus 95.42 + 273.16 = 368.58 kelvins. However, the standard deviation remains numerically the same: 2.48 kelvins. Graphically, you should picture the whole distribution moving over by the constant 273.16; the mean moves, but the standard deviation (which measures spread) doesn't change.

KEY EXAMPLE

Measurements of the sizes of farms in an upstate New York county yield a mean of 67.5 hectares with a standard deviation of 15.4 hectares. If all the measurements are converted from hectares (metric system) to acres (one acre was originally the area a yoke of oxen could plow in one day), what are the new mean and standard deviation?

Answer: One hectare is equivalent to 2.471 acres. The new mean is thus 2.471 × 67.5 = 166.8 acres with a standard deviation of 2.471 × 15.4 = 38.1 acres. Graphically, multiplying each value by the constant 2.471 both moves and spreads out the distribution.

Key 4 Five-number summary, boxplots, and outliers

OVERVIEW *The five-number summary consists of the smallest value, the largest value, the middle (median), the middle of the bottom half of the set (Q_1), and the middle of the top half of the set (Q_3). A boxplot (also called a box and whisker display) is a visual representation of dispersion that shows these five numbers.*

KEY EXAMPLE

The total farm product indexes for all years from 1919 through 1945 (with 1910–1914 as 100) are as follows: 215, 210, 130, 140, 150, 150, 160, 150, 140, 150, 150, 125, 85, 70, 75, 90, 115, 120, 125, 100, 95, 100, 130, 160, 200, 200, 210. (Note the instability of prices received by farmers!) In these data, the largest value is 215, the smallest is 70, the median is 140, the median of the top half is 160, and the median of the bottom half is 100. A boxplot of these five numbers is as follows:

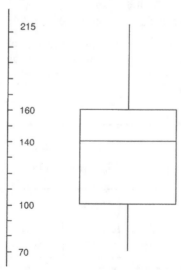

Note that the display consists of two "boxes" together with two "whiskers"—hence the alternative name. The boxes show the spread of the two middle quarters, while the whiskers show the spread of the two outer quarters. This relatively simple display conveys information not immediately available from histograms (Key 10) or stemplots (Key 16).

Note also that putting the above data into a list, for example, L1, on the TI-83, not only gives the five-number summary

```
1-Var Stats

MinX=70

Q1=100

Med=140

Q3=160

MaxX=215
```

but also gives the boxplot itself using STAT PLOT, choosing the boxplot from among the six type choices, and then using ZoomStat or in WINDOW letting Xmin = 0 and Xmax = 225.

The interquartile range (IQR) can be calculated quickly from two of the numbers in the five-number summary: $IQR = Q_3 - Q_1$. Thus the boxes in the boxplot show the spread of the IQR.

A commonly accepted definition of outliers is all values more than $1.5 \times IQR$ below Q_1 or more than $1.5 \times IQR$ above Q_3. Sometimes these outliers are plotted separately in boxplots. (The TI-83 has a modified boxplot option.)

KEY EXAMPLE

Suppose the following table summarizes a set of values.

Values below Q_1	Q_1	Median	Q_3	Values above Q_3
43,47,51	52	55	56	60,65,68

We calculate IQR = 56 – 52 = 4 and 1.5 × IQR = 6.

$Q_1 - 6 = 52 - 6 = 46$ and so 43 is an outlier.

$Q_3 + 6 = 56 + 6 = 62$ and so 65 and 68 also are outliers.

The modified boxplot (showing outliers) is

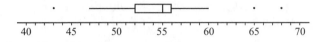

Note that the whiskers are drawn to the lowest and highest values that are not outliers.

It should be noted that two sets can have the same five-number summary and thus the same boxplots but have dramatically different distributions.

KEY EXAMPLE

Let A = {0, 5, 10, 15, 25, 30, 35, 40, 45, 50, 71, 72, 73, 74, 75, 76, 77, 78, 100} and B = {0, 22, 23, 24, 25, 26, 27, 28, 29, 50, 55, 60, 65, 70, 75, 85, 90, 95, 100}. Simple inspection indicates very different distributions; however the TI-83 gives identical boxplots with Min = 0, $Q_1 = 25$, Med = 50, $Q_3 = 75$, and Max = 100 for each.

Key 5 Position

OVERVIEW *We have seen several ways of choosing a value to represent the center of a distribution. It is also important to be able to talk about the position of any other value. In some situations, such as wine tasting, one is interested in simple rankings. Other cases, for example, evaluating college applications, may involve positioning according to percentile rankings. There are also situations in which we are able to specify position by making use of measurements of both central tendency and variability, that is, by the z-score.*

Simple ranking: Simple ranking, which involves arranging the elements in some order and noting where in the order a particular value falls, is straightforward. We know what it means for someone to graduate second in a class of 435, or for a player from a team of size 30 to have the seventh-best batting average. Simple ranking is useful even when no numerical values are associated with each element. For example, detergents may be ranked according to relative cleansing ability without any numerical measurements of strength.

Percentile ranking: Percentile ranking, which indicates what percent of all scores fall below the value under consideration, is helpful for comparing positions with different bases. We can easily compare a rank of 176 out of 704 with a rank of 187 out of 935 by noting that the first equals a percentile rank of 75%, the second a rank of 80%. Percentile rank is also useful when the exact population size is not known or is irrelevant. For example, it is more meaningful to say that a student scored in the 90th percentile on a national exam, rather than trying to determine an exact ranking among some large number of test takers.

The z-score: The z-score is a measure of position that takes into account both the center and the dispersion of the distribution. More specifically, the z-score of a value tells how many standard deviations the value is from the mean. Mathematically, $x - \mu$ gives the raw distance from μ to x; dividing by σ converts this distance to numbers of standard deviations. Thus $z = (x - \mu)/\sigma$, where x is a raw score, μ is the mean, and σ is the standard deviation. Note that, if the score x is greater than the mean μ, then z is positive, if less, then z is negative.

Given a z-score, we can also reverse the procedure and find the corresponding raw score. Solving for x gives: $x = \mu + z\sigma$.

KEY EXAMPLE

Suppose that the average (mean) price of gasoline in a large city is $1.80 per gallon with a standard deviation of $0.05. Then $1.90 has a z-score of $(1.90 - 1.80)/0.05 = +2$, while $1.65 has a z-score of $(1.65 - 1.80)/0.05 = -3$. Alternatively, a z-score of $+2.2$ corresponds to a raw score of $1.80 + 2.2(0.05) = 1.80 + 0.11 = 1.91$, while a z-score of -1.6 corresponds to $1.80 - 1.6(0.05) = 1.72$. It is often useful to portray integer z-scores and the corresponding raw scores as follows:

1.65	1.70	1.75	1.80	1.85	1.90	1.95	price/gallon
−3	−2	−1	0	1	2	3	z-score

KEY EXAMPLE

An assembly line produces an average of 12,600 units per month with a standard deviation of 830 units. Adding and subtracting multiples of 830 to the mean 12,600 gives:

10,110	10,940	11,770	12,600	13,430	14,260	15,090	units/month
−3	−2	−1	0	1	2	3	z-score

11,106 has a z-score of $(11,106 - 12,600)/830 = -1.8$, while a z-score of 2.4 corresponds to $12,600 + 2.4(830) = 14,592$

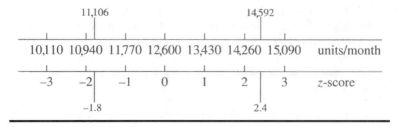

Key 6 The empirical rule

OVERVIEW *The **empirical rule** applies specifically to symmetric, "bell-shaped" data. In this case, about 68% of the values lie within one standard deviation of the mean, about 95% of the values lie within two standard deviations of the mean, and more than 99% of the values lie within three standard deviations of the mean.*

In terms of z-scores we have the following displays:

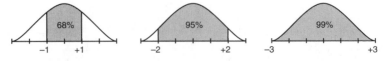

KEY EXAMPLE

Suppose that taxi cabs in New York City are driven an average of 75,000 miles per year with a standard deviation of 12,000 miles. Assuming that the distribution is "bell-shaped," we can conclude that approximately 68% of the taxis are driven between 63,000 and 87,000 miles per year, approximately 95% are driven between 51,000 and 99,000 miles, and virtually all are driven between 39,000 and 111,000 miles.

The empirical rule also gives a useful quick estimate for the standard deviation in terms of the range. We can see in the display above that 95% of the data fall within a span of four standard deviations (from −2 to +2 on the z-score line), and 99% of the data fall within a span of six standard deviations (from −3 to +3 on the z-score line). It is therefore reasonable to conclude that the standard deviation is roughly between one-fourth and one-sixth of the range. Since we can find the range of a set almost immediately, this technique for estimating the standard deviation is often helpful also in pointing out probable arithmetic errors.

KEY EXAMPLE

If the range of a data set is 60, the standard deviation should be expected to be between $(1/6)60 = 10$ and $(1/4)60 = 15$. If the standard deviation is calculated to be 0.32 or 87, there is a probable arithmetic error; a calculation of 12, however, is reasonable.

Key 7   Chebyshev's theorem

OVERVIEW   *When data are spread out, the standard deviation is larger; when data are tightly compacted, the standard deviation is smaller. However, no matter what the dispersion, and even if the data are not "bell-shaped," certain percentages of the data will always fall within specified numbers of standard deviations from the mean.*

The Russian mathematician Chebyshev showed that, for any set of data, *at least* $1 - 1/k^2$ of the values lie within k standard deviations of the mean. As a percent, at least $100(1 - 1/k^2)\%$ of the values are within k standard deviations of the mean. In terms of z-scores, at least $1 - 1/k^2$ of the values have z-scores between $-k$ and $+k$. Therefore, when $k = 3$, at least $1 - 1/9$ or 88.89% of the values lie within three standard deviations of the mean. And, when $k = 5$, at least $1 - 1/25$ or 96% of the values have z-scores between -5 and $+5$.

KEY EXAMPLE

Suppose an electronic part takes an average of 3.4 hours to move through an assembly line with a standard deviation of 0.5 hour. Then, using $k = 2$, we find that at least $1 - 1/4$ or 75% of the parts take between $3.4 - 2(0.5) = 2.4$ hours and $3.4 + 2(0.5) = 4.4$ hours to move through the line. Similarly, with $k = 4$ at least 15/16 or 93.75% of the parts take between 1.4 and 5.4 hours.

Given a range around the mean, we can convert to z-scores and ask about the percentage of values in the range. If one set has a mean $\mu = 85$ and standard deviation $\sigma = 1$, while a second set has $\mu = 85$ and $\sigma = 5$, then, for example, the percentages of values between 75 and 95 will be different. For the first set, the relevant z-scores are ± 10, while for the second the relevant z-scores are ± 2. Thus 99/100 or 99% of the first set's values, but only 3/4 or 75% of the second set's values, lie between 75 and 95.

KEY EXAMPLE

Suppose the daily intake at a tollbooth averages $3500 with a standard deviation of $200. What percentage of the daily intakes should be between $3000 and $4000?

Answer: The relevant *z*-scores are $\pm 500/200 = \pm 2.5$, so Chebyshev's theorem says that at least $1 - 1/(2.5)^2 = 21/25$ or 84% of the daily intakes should be between $3000 and $4000.

Note that, for $k = 1$, $1 - 1/k^2 = 0$; here the theorem gives no useful information.

Reminder: The power of Chebyshev's theorem is that it applies to **all** sets of data. However, if the data are "bell-shaped," we can draw stronger conclusions by using the empirical rule, stated in Key 6.

KEY EXAMPLE

Shape will be discussed in Theme 2, but intuitively consider the following two sets of data:

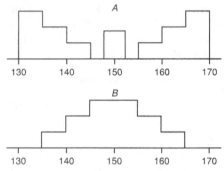

True or false?

Both sets have about the same mean.
The variance of set *A* is greater than the variance of set *B*.
Chebyshev's theorem applies to both sets.
The empirical rule applies only to set *A*.
You can be sure that the standard deviation of set *A* is greater than 5.

Answers: T (both are about 150), T, T (Chebyshev's theorem applies to all sets), F (the empirical rule applies only to bell-shaped data like set *B*), T (by Chebyshev's theorem at least 75% of the data is within two standard deviations of the mean, but only a small portion of set *A* is between 140 and 160).

14

Key 8   Theme exercises with answers

OVERVIEW   *Sample questions of the type that might appear on homework assignments and tests are presented with answers*

- Find the median, mode, and mean of these two sets: $\{2, 8, 33, 2, 20\}$ and $\{2, 8, 58, 2, 20\}$. What principle is illustrated?
  Answer: Remember to first arrange the data in numerical order to find the median.

    For $\{2, 2, 8, 20, 33\}$ the median is 8, the mode is 2, and the mean is $(2 + 2 + 8 + 20 + 33)/5 = 13$.

    For $\{2, 2, 8, 20, 58\}$ the median is 8, the mode is 2, and the mean is $(2 + 2 + 8 + 20 + 58)/5 = 18$.

  This example illustrates that the mean, unlike the median and mode, is sensitive to any change in value.

- Calculate the mean of each of these sets: $\{3, 7, 15, 22, 23, 38\}$, $\{6, 10, 18, 25, 26, 41\}$, and $\{6, 14, 30, 44, 46, 76\}$. What principle is illustrated?
  Answer: The three resulting means are 18, $18 + 3 = 21$, and $18 \times 2 = 36$. Adding or multiplying the same constant to each value will do the same to the mean.

- What can you conclude if the standard deviation of a set of observations is zero?
  Answer: Because of the squaring operation in the definition, the standard deviation (and also the variance) can be zero only if all values in the set are the same.

- Assuming that batting averages have a bell-shaped distribution, arrange the following in ascending order:
    - I. An average with a z-score of -1.
    - II. An average with a percentile rank of 20%.
    - III. An average at the first quartile, Q_1.

  Answer: Given that the empirical rule applies, a z-score of -1 has a percentile rank of about 16%. The first quartile Q_1 has a percentile rank of 25%, so the answer is I, II, III.

- Suppose the starting salaries of a graduating class are as follows:

Number of students	Starting salary ($)
10	15,000
17	20,000
25	25,000
38	30,000
27	35,000
21	40,000
12	45,000

What is the mean starting salary?
Answer: There are a total of $10 + 17 + 25 + 38 + 27 + 21 + 12 = 150$ students. Their total salary is $10(15,000) + 17(20,000) + 25(25,000) + 38(30,000) + 27(35,000) + 21(40,000) + 12(45,000) = \$4,580,000$. The mean is $4,580,000/150 \approx \$30,533$.

- Suppose the attendance at a movie theater averages 780 with a standard deviation of 40. What z-score corresponds to an attendance of 835?
Answer: $(835 - 780)/40 = 1.375$.

  What attendance corresponds to a z-score of -2.15?
Answer: $780 - 2.15(40) = 694$.

- Estimate the standard deviation of a "bell-shaped" set of data with range 180.
Answer: By the empirical rule σ is roughly between one-sixth and one-fourth of the range, or in this example between $180/6 = 30$ and $180/4 = 45$.

- Suppose the average noise level in a restaurant is 30 decibels with a standard deviation of 4 decibels. According to Chebyshev's theorem, what percentage of the time is the noise level between 22 and 38 decibels?
Answer: 22 and 38 are each $8/4 = 2$ standard deviations from the mean, yielding $1 - 1/2^2 = 3/4 = 75\%$.

- Suppose $X = \{2, 9, 11, 22\}$ and $Y = \{5, 7, 15\}$. Form the set Z of differences by subtracting each element of Y from each of X: $Z = \{2 - 5, 2 - 7, 2 - 15, 9 - 5, 9 - 7, 9 - 15, 11 - 5, 11 - 7, 11 - 15, 22 - 5, 22 - 7, 22 - 15\} = \{-3, -5, -13, 4, 2, -6, 6, 4, -4, 17, 15, 7\}$. Calculate the means μ_x, μ_y, and μ_z, and the variances σ_x^2, σ_y^2, and σ_z^2. What principle is illustrated?
Answer: $\mu_x = 11$, $\mu_y = 9$, $\mu_z = 2$, $\sigma_x^2 = 51.5$, $\sigma_y^2 = 18.67$, and $\sigma_z^2 = 70.17$. Note that $\mu_z = \mu_x - \mu_y$ and $\sigma_z^2 = \sigma_x^2 + \sigma_y^2$. This is true in general: the mean of a set of differences is equal to the *difference* of the means of the two original sets, while the variance of a set of differences is equal to the *sum* of the variances of the two original sets.

Theme 2 SHAPE

*T*here are a variety of ways to organize and arrange data. Much information can be put into tables, but these arrays of bare figures tend to be spiritless and sometimes forbidding. Some form of graphical display is often best for seeing patterns and shapes and for giving an impression of everything at once. In particular, one should always note center, spread, shape, and unusual occurrences.

Key 9   Dotplots and bar charts

OVERVIEW   *Dotplots and bar charts are particularly useful with regard to categorical (or qualitative) variables—that is, variables that note the category to which each individual belongs. This is in contrast to quantitative variables, which take on numerical values.*

KEY EXAMPLE

Suppose that in a group of 16 teenagers, 5 choose hip-hop as their favorite music, 4 choose R&B, 3 choose country, and 4 choose rock. These data can be displayed in the following *dotplot:*

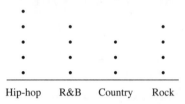

The *frequency* of each result is indicated by the *number of dots* representing that result.

KEY EXAMPLE

In a survey of college students, 25% said they play video games to relax, 40% watch TV, 20% read books, and 15% play sports. These data can be displayed in the following *bar chart:*

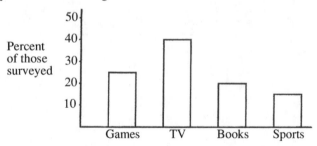

The *relative frequencies* of different results are indicated by the *heights* of the bars representing these results.

Key 10 Histograms

OVERVIEW *The **histogram** is an important visual representation of data in which relative frequencies are represented by relative areas. Useful for large data sets involving quantitative variables, it shows counts or percents falling either at certain values or between certain values.*

KEY EXAMPLE

Suppose there are 2000 families in a small town and the distribution of children among them is as follows: 300 families have no child, 400 have one child, 700 have two children, 300 have three, 100 have four, 100 have five, and 100 have six. These data can be displayed in the following histogram.

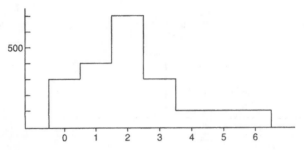

Sometimes, instead of labeling the vertical axis with *frequencies,* it is more convenient or more meaningful to use *relative frequencies,* that is, frequencies divided by the total number in the population.

Number of children	Frequency	Relative frequency
0	300	300/2000 = .150
1	400	400/2000 = .200
2	700	700/2000 = .350
3	300	300/2000 = .150
4	100	100/2000 = .050
5	100	100/2000 = .050
6	100	100/2000 = .050

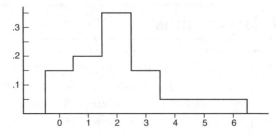

Note that the shape of the histogram is the same whether the vertical axis is labeled with frequencies or with relative frequencies.

The histogram shown above indicates the relative frequency of each value from 0 to 6, but histograms can also indicate the relative frequencies of values falling between given scores.

KEY EXAMPLE

Consider the following histogram, where the vertical axis has not been labeled:

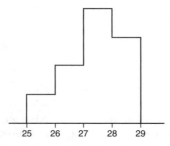

It is impossible to determine the actual frequencies; however, we can determine the relative frequencies by noting the fraction of the total *area* that is over any interval:

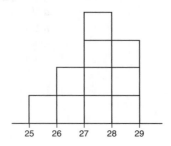

We can divide the area into ten equal portions and then note that one-tenth or 10% of the area is above 25–26, 20% is above 26–27, 40% is above 27–28, and 30% is above 28–29.

Although it will not always be possible to divide histograms so nicely into ten equal areas as above, the principle of relative frequencies corresponding to relative areas will still apply.

KEY EXAMPLE

The following histogram indicates the relative frequencies of ages of U.S. scientists in 1967.

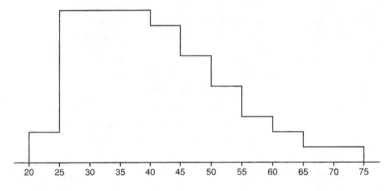

If one compares areas (or counts small rectangles!), one can conclude that 50% of the scientists are between 25 and 40 years of age, 20% are between 45 and 55 years of age, etc. Now if we also knew that there were 300,000 U.S. scientists in 1967, then we could convert these percentages to frequencies, such as 150,000 scientists between 25 and 40 years of age, or 60,000 scientists between 45 and 55 years of age.

Key 11 Reference shapes

OVERVIEW *Distributions come in an endless variety of shapes; however, certain common patterns are worth special mention.*

- A *symmetric* distribution is one in which the two halves are mirror images of each other. For example, the weights of all people in some organizations fall into symmetric distributions with two mirror bumps, one for men's weights and one for women's weights.

- A distribution is *skewed to the right* if it spreads far and thinly toward the higher values. For example, ages of nonagenarians (people in their 90s) is a distribution with sharply decreasing numbers as one moves from 90-year-olds to 99-year-olds.

- A distribution is *skewed to the left* if it spreads far and thinly toward the lower values. For example, scores on an easy exam show a distribution bunched at the higher end with few low values.

- A *bell-shaped* distribution is symmetric with a center mound and two sloping tails. For example, IQ scores across the general population is roughly symmetric with a center mound at 100 and two sloping tails.

- A distribution is *uniform* if its histogram is a horizontal line. For example, tossing a fair die and noting how many spots (pips) appear on top yields a uniform distribution with 1 through 6 all equally likely.

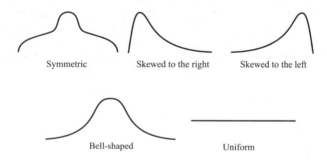

Symmetric   Skewed to the right   Skewed to the left

Bell-shaped   Uniform

Even when a basic shape is noted, it is important also to note if some of the data deviate from this shape.

Key 12 Center and spread, clusters and gaps, outliers and modes

OVERVIEW *Looking at a graphical display, some important aspects of the overall pattern to note are the **center**, which separates the values (or area under the curve in the case of a histogram) roughly in half; the **spread**, that is, the scope of the values from smallest to largest; **clusters**, which show natural subgroups into which the values fall; **gaps**, which show holes where no values fall; **outliers** (extreme values); and **modes** (major peaks that some distributions show).*

- The salaries of teachers in Ithaca, New York, fall into three overlapping *clusters*, one for public school teachers, a higher one for Ithaca College professors, and an even higher one for Cornell University professors.

- The office of the dean sends letters to students being put on the honor roll and to those being put on academic warning for low grades; thus the GPA distribution of students receiving letters from the dean has a huge middle *gap*.

- Sometimes *outliers* are the result of blunders in measurements and deserve scrutiny; however, they can also be the result of natural chance variation. It is usually instructive to look for reasons behind *outliers* and *modes*.

- If a distribution has one peak, it is said to be *unimodal;* with two major peaks, *bimodal*.

KEY EXAMPLE

Consider the following histogram:

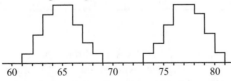

Simply saying that the *center* of the distribution is around 71 and the *spread* is from 61 to 81 clearly misses something. The distribution is *symmetric*, and its values fall into two distinct *clusters* with a *gap* between.

Key 13 Cumulative frequency plot

OVERVIEW *Sometimes we sum frequencies and show the result visually in a **cumulative frequency plot.***

KEY EXAMPLE

The following graph shows weight distribution of half-gallon milk containers coming from a bottling machine.

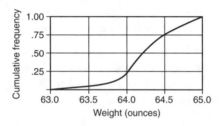

What can we learn from this cumulative frequency plot? For example, going up to the graph from weight 64.0 ounces, we see that .25 or 25% of the bottles contain less than 64.0 ounces. Going over to the graph from .50 on the vertical axis, we see that 50% of the bottles have less than 64.2 ounces, and thus 64.2 is the *median*. Going over from .25 and .75 we see that the *interquartile range* is 64.5 – 64.0 = 0.5. Since 63.0 is the minimum number of ounces and 65.0 is the maximum number of ounces, the *range* is 65.0 – 63.0 = 2.0.

A distribution skewed to the left has a cumulative frequency plot that rises slowly at first and then steeply later, whereas a distribution skewed to the right has a cumulative frequency plot that rises steeply at first and then slowly later.

KEY EXAMPLE

Consider the essay-grading policies of three teachers: Abrams, who gives very high scores; Brown, who gives equal numbers of low and high scores; and Connors, who gives very low scores. Histograms of the grades (with 1 high and 4 low) are as follows:

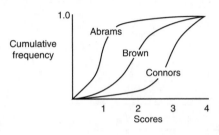

These translate into the following cumulative frequency plots:

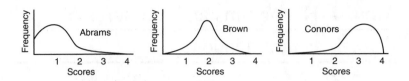

Key 14 Histograms and measures of central tendency

OVERVIEW *Suppose we have a detailed histogram such as the following:*

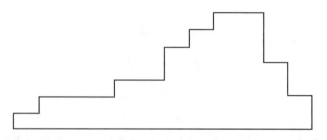

Our measures of central tendency fit naturally into such a diagram.

The *mode* is defined as the most frequent value, so it is the point or interval at which the graph is highest.

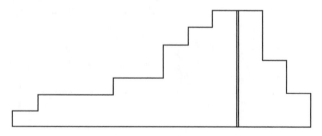

The *median* divides a distribution in half, so it is represented by a line that divides the total area of the histogram in half.

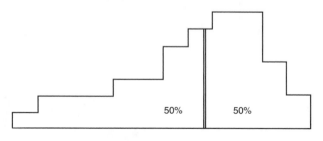

The *mean* is a value that is affected by the spacing of all the values. Therefore, if the histogram is considered to be a solid region, then the mean corresponds to a line passing through the center of gravity (where the graph would balance if it were a solid object).

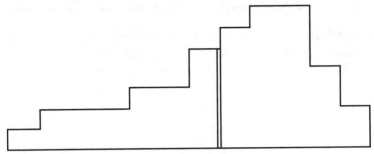

The above distribution, spread thinly far to the low side, is said to be *skewed to the left*. Note that in this case the mean is less than the median. Similarly, a distribution spread far to the high side is called *skewed to the right*, and its mean will be greater than its median.

KEY EXAMPLE

Suppose that the faculty salaries at a college have a median of $32,500 and a mean of $38,700. What does this indicate about the shape of the distribution of the salaries?

Answer: The median is less than the mean, so the salaries are probably skewed to the right—a few high paid professors with the bulk of the professors on the lower end of the pay scale.

Key 15 Histograms, z-scores, and percentile ranks

OVERVIEW *We have seen that relative frequencies are represented by relative areas and so labeling of the vertical axis is not crucial. If we know the standard deviation, the horizontal axis can be labeled in terms of z-scores. In fact, if we are given the percentile rankings of various z-scores, we can construct a histogram.*

KEY EXAMPLE

Suppose we are given the following data:

z-Score:	−2	−1	0	1	2
Percentile ranking:	0	20	60	70	100

We note that the entire area is less than z-score +2 and greater than z-score −2. Also, 20% of the area is between z-scores −2 and −1, 40% is between −1 and 0, 10% is between 0 and 1, and 30% is between 1 and 2. Thus the histogram is as follows:

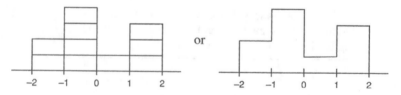

or

Suppose that we are also given four in-between z-scores:

z-Score	Percentile ranking
2.0	100
1.5	80
1.0	70
0.5	65
0.0	60
−0.5	30
−1.0	20
−1.5	5
−2.0	0

28

Then we have

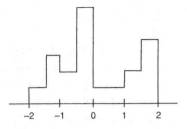

With 1000 z-scores perhaps the histogram would look like this:

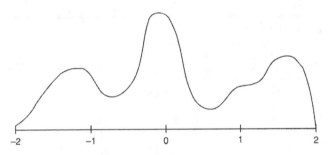

The height at any point is meaningless; what is important is relative *areas*. For example, in the final diagram, what percent of the area is between z-scores of +1 and +2?   *Answer:* Still 30%.
What percent is to the left of 0?   *Answer:* Still 60%.

Key 16 Stemplots

OVERVIEW *Although a histogram may show how many scores fall into each grouping or interval, the exact values of individual scores are often lost. An alternative pictorial display, called a stemplot (or a stem and leaf display), retains this individual information.*

KEY EXAMPLE

Consider the set $\{25, 33, 28, 31, 45, 52, 37, 31, 46, 33, 20\}$. Let 2, 3, 4, and 5 be place holders for 20, 30, 40, and 50. List the last digit of each value from the original set after the appropriate place holder. The result is the stemplot of these data:

Stems	Leaves
2	5 8 0
3	3 1 7 1 3
4	5 6
5	2

Drawing a continuous line around the leaves results in a horizontal histogram:

Note that the stem and leaf display gives the shape of the histogram. However, the histogram does not indicate the values of the original data, as does the stem and leaf display.

Sometimes further structure is shown by rearranging the numbers in each row in ascending order. The ordered display shows a second level of information from the original stemplot, and makes it easier to identify Q_1, Q_3, and the median.

KEY EXAMPLE

The revised display of the data in the above example is as follows:

2	0 5 8
3	1 1 3 3 7
4	5 6
5	2

The stems in stem and leaf displays may be other than single digits.

KEY EXAMPLE

Suppose the distribution of 25 advertised house prices (in $1000's) in a certain community is given by {56, 89, 165, 73, 83, 145, 90, 189, 127, 77, 110, 112, 132, 120, 94, 130, 84, 65, 99, 154, 86, 120, 122, 103, 130}. One possible stem and leaf display of this data is as follows:

50–74	56 73 65
75–99	89 83 90 77 94 84 99 86
100–124	10 12 20 20 03
125–149	45 27 32 30 22 30
150–174	65 54
175–200	89

Note that the stems above were chosen to be intervals of length 25, and that the "1" of the 100 is left out of the leaves so that the leaves align vertically.

Key 17 Theme exercises with answers

OVERVIEW *Sample questions of the type that might appear on homework assignments and tests are presented with answers.*

- Suppose the 40 top-level executives of a large company receive salaries (in $1000's) distributed as follows: 1 between 20 and 30, 5 between 30 and 40, 10 between 40 and 50, 12 between 50 and 60, 6 between 60 and 70, 4 between 70 and 80, and 2 between 80 and 90. Draw a histogram with two vertical axes, one showing frequencies and the other showing relative frequencies.

Answer:

- Consider the following annual murder rates (per 100,000 people) for each of the 50 states: AL—13.3, AK—12.9, AZ—9.4, AR—9.1, CA—11.7, CO—7.3, CT—4.2, DE—6.7, FL—11.0, GA—14.4, HI—6.7, ID—5.4, IL—9.9, IN—6.2, IA—2.6, KS—5.7, KY—9.0, LA—15.8, ME—2.7, MD—8.2, MA—3.7, MI—10.6, MN—2.0, MS—12.6, MO—10.4, MT—4.8, NE—3.0, NV—15.5, NH—1.4, NJ—5.4, NM—10.2, NY—10.3, NC—10.8, ND—1.2, OH—6.9, OK—8.5, OR—5.0, PA—6.2, RI—4.0, SC—11.5, SD—1.9, TN—9.4, TX—14.2, UT—3.7, VT—3.3, VA—8.8, WA—4.6, WV—6.8, WI—2.5, WY—7.1. Draw a histogram with two horizontal axes, one corresponding to raw scores and one to z-scores.

Answer: Using our basic formulas gives $\mu = 7.57$ and $\sigma = 3.91$. We calculate raw scores corresponding to various z-scores by using $x = 7.57 + 3.91z$. Finally, we count elements (2% per element) to obtain percentile rankings.

z-Score:	−2	−1.5	−1	−0.5	0	1	1.5	2	2.5	3
Raw score:	−0.25	1.71	3.66	5.62	7.57	9.53	11.48	13.43	15.39	17.34
Percentile ranking:	0	4	18	36	54	68	82	92	96	100

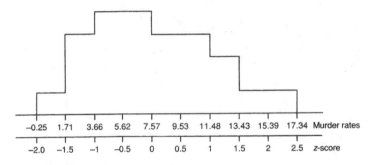

- Give a stemplot for the set {126, 195, 149, 122, 189, 164, 228, 177, 165, 150, 169, 127, 176, 147, 148, 159, 128, 122, 150, 193, 207, 164, 168, 110, 155, 127, 152, 174, 190, 219, 125, 193, 141, 127, 155, 133, 150, 162, 168, 128, 125, 137, 146, 120, 154, 176, 166, 117, 154, 137} representing the 1988 per capita personal income (in $100's) for the 50 states.

Answer:

11	0 7
12	6 2 7 8 2 7 5 7 8 5 0
13	3 7 7
14	9 7 8 1 6
15	0 9 0 5 2 5 0 4 4
16	4 5 9 4 8 2 8 6
17	7 6 4 6
18	9
19	5 3 0 3
20	7
21	9
22	8

- Draw a boxplot of the above data.

   Answer: Arranging the set in numerical order gives {110, 117, 120, 122, 122, 125, 125, 126, 127, 127, 127, 128, 128, 133, 137, 137, 141, 146, 147, 148, 149, 150, 150, 150, 152, 154, 154, 155, 155, 159, 162, 164, 164, 165, 166, 168, 168, 169, 174, 176, 176, 177, 189, 190, 193, 193, 195, 207, 219, 228}. The smallest value is 110; the largest, 228. The median is 153; the median of the top half, 169; the median of the bottom, 128.

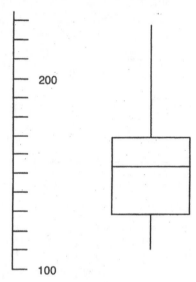

- Consider the following five boxplots.

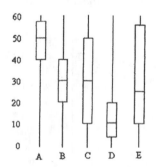

   Which of the above boxplots do each of the following histograms correspond to?

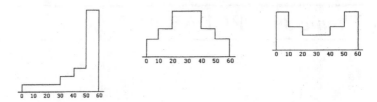

Answer: In the first histogram, the value 50 seems to split the area under the histogram in two, so the median is about 50. Furthermore, the histogram is skewed to the left with a tail from 0 to 30. Thus it corresponds to boxplot A. For the second histogram, looking at areas under the curve, Q_1 appears to be around 20, the median is around 30, and Q_3 is about 40. Thus it corresponds to boxplot B. For the third histogram, looking at areas under the curve, Q_1 appears to be around 10, the median is around 30, and Q_3 is about 50. Thus it corresponds to boxplot C.

Theme 3 PROBABILITY

*I*n the world around us, sometimes unlikely events take place, and at other times very feasible events do not occur. Because of the myriad and minute origins of various happenings, it is often impracticable, or simply impossible, to predict exact outcomes. However, while we may not be able to foretell a specific result, we may be able to assign what is called a *probability* to the likelihood of any particular event happening.

For our study of statistics, we need an understanding of the probability of an elementary event that could happen many times, each time under the same circumstances. We want to be able to deduce the chance or prospect of occurrence of such events, and then use this information to make inferences about more complex circumstances where complete information is not known. For example, we might analyze the past movements of various stock prices given various economic conditions, calculate specific probabilities, and then ask what can be said, with what degree of confidence, about future movements.

In this theme, we limit our discussion to the development of the specific techniques necessary to appreciate the basic principles of statistical analysis, which will later be considered. In particular, we need to understand what are called *binomial probabilities* and be able to calculate *expected values* regarding outcomes associated with such binomial probabilities.

INDIVIDUAL KEYS IN THIS THEME

Key 18 Elementary concepts

OVERVIEW *The **probability** of an outcome of some experiment is a mathematical statement about the likelihood of that event occurring. Probabilities are always between 0 and 1, with a probability close to 0 meaning that an event is unlikely to occur, and a probability close to 1 meaning that the event is likely to occur. The sum of the probabilities of all the separate outcomes of an experiment is always 1. In this key we list basic probability concepts that are needed for our discussion of probability distributions.*

KEY EXAMPLE

Suppose A, B, C, D, and E are events, and you calculate their probabilities to be P(A) = .5, P(B) = 0, P(C) = 1, P(D) = –0.34, and P(E) = 1.6. How would you interpret these results?

Answer: Event A is as likely to happen as not to happen, event B cannot happen, and event C is sure to happen. There is some error in your calculations of the probabilities of events D and E because probabilities are never negative and are never greater than one.

Complementary events: The probability that an event does not occur, that is, the probability of its complement, is equal to 1 minus the probability that the event does occur.

Notation: $P(X') = 1 – P(X)$

More notation: Often p symbolizes the probability of an event and q symbolizes the probability of its complement. Then $p + q = 1$.

KEY EXAMPLE

If the probability that a company will win a contract is .3, then the probability it will not win the contract is $1 – .3 = .7$.

Addition principle: If two events are mutually exclusive, that is, if they cannot occur simultaneously, then the probability of at least one occurring is equal to the sum of their respective probabilities.

If $P(X \cap Y) = 0$, then $P(X \cup Y) = P(X) + P(Y)$ (where $X \cap Y$, read as "X intersect Y," means that both X and Y occur, while $X \cup Y$, read as "X union Y," means that either X or Y occurs or both occur).

KEY EXAMPLE

If the probabilities of Jane, Lisa, and Mary being chosen chairwoman of the board are .5, .3, and .2, respectively, then the probability that the chairwoman will be either Jane or Mary is .5 + .2 = .7.

The above addition principle can be extended to more than two events. That is, the probability that any one of several mutually exclusive events occurs is equal to the sum of their individual probabilities.

General addition rule: For any pair of events X and Y,
$P(X \cup Y) = P(X) + P(Y) - P(X \cap Y)$

KEY EXAMPLE

Suppose the probability that a construction company will be awarded a certain contract is .34, the probability that it will be awarded a second contract is .27, and the probability that it will get both contracts is .18. Then the probability that the company will win at least one of the two contracts is .34 + .27 − .18 = .43.

Independence principle: If the chance of one event happening is not influenced by whether or not a second event happens, then the probability that *both* events will happen is the product of their separate probabilities.

KEY EXAMPLE

The probability that a company will receive a grant from a private concern is 1/3, while the probability that it will receive a federal grant is 1/2. If whether or not the company receives one grant is not influenced by whether or not it receives the other, then the probability of receiving both grants is $1/3 \times 1/2 = 1/6$.

The above independence principle can also be extended to more than two events. That is, given a sequence of independent events, the probability that *all* happen is equal to the product of their individual probabilities.

Conditional probability formula: The probability of the event X occurring given that the event Y has occurred is

$$P(X|Y) = \frac{P(X \cap Y)}{P(Y)}$$

KEY EXAMPLE

If 75% of the households in a certain region have answering machines and 50% have both answering machines and call waiting, then the probability that a household chosen at random and found to have an answering machine also has call waiting is calculated:

$$P\left(\begin{array}{c}call \\ waiting\end{array} \middle| \begin{array}{c}answering \\ machine\end{array}\right) = \frac{P\left(\begin{array}{c}call \\ waiting\end{array} \cap \begin{array}{c}answering \\ machine\end{array}\right)}{P\left(\begin{array}{c}answering \\ machine\end{array}\right)} = \frac{.50}{.75} = \frac{2}{3}$$

Note that we can summarize with the formulas:

In general, $P(X \cap Y) = P(X)P(Y|X)$.

However, $P(X \cap Y) = P(X)P(Y)$ if X and Y are independent.

Key 19 Multistage probability calculations

OVERVIEW *The elementary concepts presented in Key 18 can be integrated into powerful tools to solve a range of probability problems. One particular multistage calculation, called **Bayes Theorem,** is presented here.*

KEY EXAMPLE

The Air Force receives 30% of its parachutes from company C1 and the rest from company C2. The probability that a parachute will fail to open is .0025 and .002, depending on whether it is from company C1 or C2, respectively. What is the probability that a randomly chosen parachute will fail to open?

Answer: In such problems it is helpful to draw a tree diagram like the following:

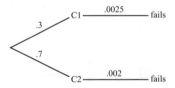

We then have

$$P(C1 \cap fails) = (.3)(.0025) = .00075$$
$$P(C2 \cap fails) = (.7)(.002) = .0014$$

At this stage a Venn diagram is helpful in finishing the problem:

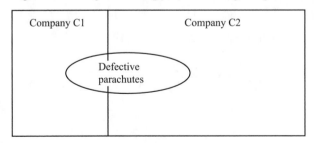

$$P(\text{fails}) = P(\text{C1} \cap \text{fails}) + P(\text{C2} \cap \text{fails}) = .00075 + .0014 = .00215$$

We can take the above analysis one stage further and answer such questions as these: If a randomly chosen parachute fails, what is the probability it came from company C1? From company C2?

$$P(\text{C1}|\text{fails}) = \frac{P(\text{C1} \cap \text{fails})}{P(\text{fails})} = \frac{.00075}{.00215} = .349$$

$$P(\text{C2}|\text{fails}) = \frac{P(\text{C2} \cap \text{fails})}{P(\text{fails})} = \frac{.0014}{.00215} = .651$$

KEY EXAMPLE

A travel agent books passages on three different tours, with half her customers choosing tour T1, one-third choosing T2, and the rest choosing T3. The agent has noted that three-quarters of those who take tour T1 return to book passage again, two-thirds of those who take T2 return, and one-half of those who take T3 return. If a customer does return, what is the probability that the person first went on tour T2?

Answer: $P(\text{T3}) = 1 - (1/2 + 1/3) = 1/6$.

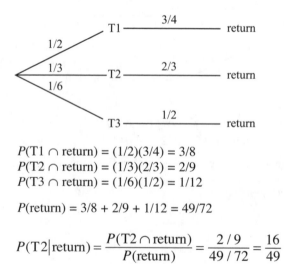

$P(\text{T1} \cap \text{return}) = (1/2)(3/4) = 3/8$
$P(\text{T2} \cap \text{return}) = (1/3)(2/3) = 2/9$
$P(\text{T3} \cap \text{return}) = (1/6)(1/2) = 1/12$

$P(\text{return}) = 3/8 + 2/9 + 1/12 = 49/72$

$$P(\text{T2}|\text{return}) = \frac{P(\text{T2} \cap \text{return})}{P(\text{return})} = \frac{2/9}{49/72} = \frac{16}{49}$$

Key 20 Binomial formula

OVERVIEW *In many applications there are two possible outcomes. For example, either a radio is defective or it is not defective, either the workers will go on strike or they won't go on strike, either the manager's salary is above $50,000 or it is not above $50,000. For applications in which a two-outcome situation is repeated some number of times, and the probability of each of the two outcomes remains the same for each repetition, the resulting calculations involve what are known as **binomial probabilities**.*

Notation: In the discussion below, $n!$ (read as "n factorial") denotes the product of all the integers from n down to 1; that is, $n! = n(n-1)(n-2) \ldots (3)(2)(1)$.

KEY EXAMPLE

Suppose that 30% of the employees in a large factory are smokers. What is the probability that there will be exactly two smokers in a randomly chosen five-person work group?

Answer: We reason as follows. The probability that a person smokes is 30% = .3, so the probability that he or she doesn't smoke is $1 - .3 = .7$. The probability of a particular arrangement of two smokers and three nonsmokers is $(.3)^2(.7)^3 = .03087$. The number of such arrangements is given by the *combination* $C(5,2) = 5!/2!3! = 10$. Each such arrangement has probability .03087, so the final answer is $10 \times .03087 = .3087$. [On the TI-83, binompdf(5,.3,2) = .3087.]

Binomial formula: Suppose an experiment has two possible outcomes, called *success* and *failure,* with the probability of success equal to p and the probability of failure equal to q (of course, $p + q = 1$). Suppose further that the experiment is repeated n times, and the outcome at any particular time does not have any influence over the outcome at any other time. Then the probability of exactly x successes (and thus $n - x$ failures) is

$$C(n,x)p^x q^{n-x} = \frac{n!}{x!(n-x)!}p^x q^{n-x}$$

KEY EXAMPLE

A manager notes that there is a .125 probability that any employee will arrive late for work. What is the probability that exactly one person in a six-person department will arrive late?

Answer: If the probability of being late is .125, then the probability of being on time is $1 - .125 = .875$. If one person out of six is late, then $6 - 1 = 5$ will be on time. $C(6,1) = 6!/5!1! = 6$. Thus the desired probability is $C(6,1)(.125)^1(.875)^5 = 6(.125)(.875)^5 = .385$. [Or binompdf(6,.125,1) = .385.]

Many, perhaps most, applications involve such phrases as *at least, at most, less than, more than*. In these cases solutions involve summing together two or more cases.

KEY EXAMPLE

A manufacturer has the following quality-control check at the end of a production line: If at least eight out of ten randomly picked articles meet all specifications, the whole shipment is approved. If, in reality, 85% of the shipment would meet all specifications, what is the probability that the shipment will make it through the control check?

Answer: The probability of meeting specifications is .85, so the probability of not meeting specifications must be .15. We want the probability that at least eight out of ten articles meet specifications, that is, the probability that exactly eight or exactly nine or exactly ten meet specifications. We sum the three binomial probabilities:

Exactly 8 out of 10 meet specifications	Exactly 9 out of 10 meet specifications	Exactly 10 out of 10 meet specifications

$C(10,8)(.85)^8(.15)^2 + C(10,9)(.85)^9(.151 + C(10,10)(.85)^{10}(.15)^0$

$$= \frac{10!}{8!2!}(.85)^8(.15)^2 + 10(.85)^9(.15) + (.85)^{10} = .820$$

[On the TI-83 one can calculate $1 - $ binomcdf(10,.85,7) = .820 or binomcdf(10,.15,2) = .820.]

In some situations it is easier to calculate the probability of the complimentary event and subtract this value from 1.

KEY EXAMPLE

Joe DiMaggio had a career batting average of .325. What was the probability that he would get at least one hit in five official times at bat?

Answer: We could sum the probabilities of exactly one hit, two hits, three hits, four hits, and five hits. However, the complement of "at least one hit" is "zero hit." The probability of no hit is $C(5,0)(.325)^0(.675)^5 = (.675)^5 = .140$, and thus the probability of at least one hit in five times at bat is $1 - .140 = .860$. [On the TI-83 either $1 - binompdf(5,.325,0) = .860$ or $binomcdf(5,.675,4) = .860$.]

Sometimes we are asked to calculate the probability of each of the separate outcomes (these should sum to 1).

KEY EXAMPLE

If the probability of a male birth is .51, what is the probability that a five-child family will have all boys? Exactly four boys? Exactly three boys? Exactly two boys? Exactly one boy? All girls?

Answer:

$$
\begin{aligned}
P(5 \text{ boys}) &= C(5,5)(.51)^5(.49)^0 = (.51)^5 & = & \quad .0345 \\
P(4 \text{ boys}) &= C(5,4)(.51)^4(.49)^1 = 5(.51)^4(.49) & = & \quad .1657 \\
P(3 \text{ boys}) &= C(5,3)(.51)^3(.49)^2 = 10(.51)^3(.49)^2 & = & \quad .3185 \\
P(2 \text{ boys}) &= C(5,2)(.51)^2(.49)^3 = 10(.51)^2(.49)^3 & = & \quad .3060 \\
P(1 \text{ boy}) &= C(5,1)(.51)^1(.49)^4 = 5(.51)(.49)^4 & = & \quad .1470 \\
P(0 \text{ boys}) &= C(5,0)(.51)^0(.49)^5 = \quad (.49)^5 & = & \quad \underline{.0283} \\
& & & 1.0000
\end{aligned}
$$

[Or $binompdf(5,.49) = \{.0345\ .1657\ .3185\ .3060\ .1470\ .0283\}$.]

Note: In calculating *combinations,* 0! Is defined to be 1; thus $C(5,5) = 5!/5!0! = 1$, and $C(5,0) = 5!/0!5! = 1$.

The binomial probability of x successes can be found from the probability of $x - 1$ successes using the formula:

$$
P(x \text{ successes}) = \frac{n-x+1}{x}\frac{p}{1-p}P(x-1 \text{ successes})
$$

KEY EXAMPLE

A marksman can hit a bull's-eye target 95% of the time. Given that the probability of exactly 8 bulls-eyes in 10 shots is .0746, what is the probability of exactly 9 bulls-eyes in 10 shots?

Answer:

$$\frac{10 - 9 + 1}{9} \frac{.95}{1 - .95} (.0746) = .315$$

Key 21 Random variables

OVERVIEW *Often each outcome of an experiment has not only an associated probability, but also an associated **real number**. For example, the probability might be 1/2 that there are five defective batteries; the probability might be .01 that a company will receive seven contracts; the probability might be .95 that three people will recover from a disease. If* X *represents the different numbers associated with the potential outcomes of some situation, then we call* X *a **random variable**.*

KEY EXAMPLE

A town prison official knows that 1/2 the inmates he admits stay only 1 day, 1/4 stay 2 days, 1/5 stay 3 days, and 1/20 stay 4 days before they are either released or sent on to the county jail. If X represents the number of days, then X is a *random variable* that takes the values 1, 2, 3, and 4. X takes the value 1 with probability 1/2, the value 2 with probability 1/4, the value 3 with probability 1/5, and the value 4 with probability 1/20.

The random variable in the example above is called **discrete** because it can assume only a countable number of values, while the one in the following example is called **continuous** because it can assume values associated with a whole line interval.

KEY EXAMPLE

Let X be a random variable whose values correspond to the speeds at which a jet plane can fly. Note that the jet might be traveling at 623.478 … mph or any other value in some whole interval. We might ask what the probability is that the plane is flying between 300 and 400 mph.

A **probability distribution** for a discrete variable is a listing or formula giving the probability for each value of the random variable.

KEY EXAMPLE

Concessionaires know that attendance at a football stadium will be 60,000 on a clear day, 45,000 if there is light snow, and 15,000 if there is heavy snow. Furthermore, the probability of clear skies, light snow, or heavy snow on any particular day is 1/2, 1/3, or 1/6, respectively. (Here we have a random variable X that takes the values 60,000, 45,000, and 15,000.)

Outcome:	Clear skies	Light snow	Heavy snow
Probability:	1/2	1/3	1/6
Random variable:	60,000	45,000	15,000

If the probabilities come from the binomial formula, then we have what is called a **binomial probability distribution.**

Key 22 Expected value or mean of a random variable

OVERVIEW *The **expected value** (or **average** or **mean**) of a random variable* X *(with a finite number of values) is the sum of the products obtained by multiplying each value* x *by the corresponding probability* P(x). *We write:*

$$E(X) = \sum x \, P(x)$$

KEY EXAMPLE

In a lottery, 10,000 tickets are sold with a prize of $7500 for the one winner. The actual winning payoff is $7499 because the winner paid for his $1 ticket, so we have:

Outcome:	Win	Lose
Probability	$\dfrac{1}{10,000}$	$\dfrac{9999}{10,000}$
Random variable:	7499	−1

The expected value is $7499\left(\dfrac{1}{10,000}\right) + (-1)\left(\dfrac{9999}{10,000}\right) = -0.25$.

Thus the *average* result for each person betting the lottery is a 25¢ loss.

KEY EXAMPLE

A manager must choose among three options. Option A has a 10% chance of resulting in a $250,000 gain, but otherwise will result in a $10,000 loss. Option B has a 50% chance of gaining $40,000 and a 50% chance of losing $2000. Finally, option C has a 5% chance of gaining $800,000, but otherwise will result in a loss of $20,000. Which option should the manager choose?

Answer:

	Option A		Option B		Option C	
	Gain	Loss	Gain	Loss	Gain	Loss
Outcome:						
Probability:	.10	.90	.50	.50	.05	.95
Random variable:	250,000	−10,000	40,000	−2000	800,000	−20,000

$$E(A) = .10(250.000) + .90(-10,000) = \$16,000$$
$$E(B) = .50(40.000) + .50(-2000) = \$19,000$$
$$E(C) = .05(800.000) + .95(-20,000) = \$21,000$$

The manager should choose option C!

Although option C has the greatest mean, the manager might well wish to consider the relative riskiness of each option. If, for example, a $5000 loss would be disastrous for the company, the manager might well decide to choose option B with its maximum possible loss of $2000. In the next key, we will consider how to measure the *variability* of a random variable.

KEY EXAMPLE

One investment has two possible returns: $3000 with a probability 1/4 and $2000 with probability 3/4. A second investment has possible returns of $6000, $7000, and $9000 with probabilities of 1/6, 1/2, and 1/3, respectively. Assume that what happens on one investment is independent of what happens on the other. What are the expected values for the return on each investment and on the total investment?

Answer:

$$E(X) = 3000(1/4) + 2000(3/4) = \$2250$$

$$E(Y) = 6000(1/6) + 7000(1/2) + 9000(1/3) = \$7500$$

For the total investment Z we have

x	$P(x)$		x	$P(x)$
3000 + 6000	(1/4)(1/6)		9000	1/24
3000 + 7000	(1/4)(1/2)		10000	1/8
3000 + 9000	(1/4)(1/3)	or	12000	1/12
2000 + 6000	(3/4)(1/6)		8000	1/8
2000 + 7000	(3/4)(1/2)		9000	3/8
2000 + 9000	(3/4)(1/3)		11000	1/4

$$E(Z) = 9000(1/24) + \cdots + 11000(1/4) = 9750$$

Note that $E(X) + E(Y) = E(Z)$.

Key 23 Variance and standard deviation

of a random variable

OVERVIEW   *Not only is the mean important, but also we would like to measure the **variability** for the values taken on by a random variable. We are dealing with chance events, so the proper tool is **variance.***

In Key 2 variance was defined as the mean average of the squared deviations $(x - \mu)^2$. If we regard the $(x - \mu)^2$ terms as the values of some random variable (whose probability is the same as the probability of x), then the mean of this new random variable is simply $\Sigma(x - \mu)^2 P(x)$. This is precisely how we define the variance σ^2 of a discrete random variable:

$$\sigma^2 = \Sigma(x - \mu)^2 P(x)$$

KEY EXAMPLE

A highway engineer knows that his workers can lay 5 miles of highway on a clear day, 2 miles on a rainy day, and only 1 mile on a snowy day. Suppose the probabilities are as follows:

Outcome:	Clear	Rain	Snow
Probability:	.6	.3	.1
Random variable (miles of highway):	5	2	1

Then the mean, or expected value, and the variance are calculated as follows:

$$\mu = \Sigma x\, P(x) = 5(.6) + 2(.3) + 1(.1) = 3.7$$

$$\sigma^2 = \Sigma(x - \mu)^2\, P(x) = (5 - 3.7)^2\, (.6) + (2 - 3.7)^2(.3) + (1 - 3.7)^2(.1)$$
$$= 2.61$$

As before, the standard deviation σ is the square root of the variance.

A computational tool giving the same results is as follows:

$$\sigma^2 = \Sigma(x - \mu)^2\, P(x) = \Sigma x^2 P(x) - \mu^2$$

KEY EXAMPLE

A particular stock investment will yield the following profit per share with the given probability:

Profit ($):	0	5	10	15	20
Probability:	.3	.3	.2	.1	.1

Then

$$\mu = 5(.3) + 10(.2) + 15(.1) + 20(.1) = 7$$
$$\sigma^2 = 25(.3) + 100(.2) + 225(.1) + 400(.1) - 49 = 41$$
$$\sigma = \sqrt{41} = 6.403$$

KEY EXAMPLE

For the last example of Key 22 on page 50, what are the three variances, and what point is illustrated?

Answer:

$$\sigma_X^2 = (3000)^2(1/4) + (2000)^2(3/4) - (2250)^2 = 187{,}500$$
$$\sigma_Y^2 = (6000)^2(1/6) + (7000)^2(1/2) + (9000)^2(1/3) - (7500)^2 = 1{,}250{,}000$$
$$\sigma_Z^2 = (9000)^2(1/24) + (10{,}000)^2(1/8) + (12{,}000)^2(1/12) + (8000)^2(1/8)$$
$$+ (9000)^2(3/8) + (11{,}000)^2(1/4) - (9750)^2 = 1{,}437{,}500$$

Note that $\sigma_X^2 + \sigma_Y^2 = \sigma_Z^2$

Key 24 Mean and standard deviation

of a binomial

OVERVIEW *In the case of a **binomial random variable**, that is, a random variable whose values are the number of "successes" in some binomial probability distribution, there is a shortcut to calculate the mean and standard deviation.*

KEY EXAMPLE

Of the automobiles produced in a particular plant, 40% have a certain defect. Suppose a company purchases five of these cars. What is the expected value for the number of cars with defects?

Answer: We might guess that the average or mean or expected value is 40% of 5 = .4 × 5 = 2, but let us calculate from the definition. Letting X represent the number of cars with the defect, we have:

$$P(0) = C(5,0)(.4)^0(.6)^5 = (.6)^5 = .07776$$
$$P(1) = C(5,1)(.4)^1(.6)^4 = 5(.4)(.6)^4 = .25920$$
$$P(2) = C(5,2)(.4)^2(.6)^3 = 10(.4)^2(.6)^3 = .34560$$
$$P(3) = C(5,3)(.4)^3(.6)^2 = 10(.4)^3(.6)^2 = .23040$$
$$P(4) = C(5,4)(.4)^4(.6)^1 = 5(.4)^4(.6) = .07680$$
$$P(5) = C(5,5)(.4)^5(.6)^0 = (.4)^5 = .01024$$

Outcome:	0 car	1 car	2 cars	3 cars	4 cars	5 cars
Probability:	.07776	.25920	.34560	.23040	.07680	.01024
Random variable:	0	1	2	3	4	5

$$E(X) = 0(.07776) + 1(.25920) + 2(.34560) + 3(.23040)$$
$$+ 4(.07680) + 5(.01024) = 2$$

Thus, the answer turns out to be the same as would be obtained by simply multiplying the probability of "success" times the number of cases.

The following is true: If we have a binomial probability situation with the probability of success equal to p and the number of trials equal to n, then the *expected value* or *mean* number of successes for the n trials is np.

KEY EXAMPLE

An insurance salesperson is able to sell policies to 15% of the people she contacts. Suppose she contacts 120 people during a 2-week period. What is the expected value for the number of policies she sells?

Answer: We have a binomial probability with the probability of success .15 and the number of trials 120, so the mean or expected value for the number of successes is $120 \times .15 = 18$.

Similarly, for this case there is a shortcut calculation of the variance and standard deviation. We have

$$\sigma^2 = np(1-p) \text{ and } \sigma = \sqrt{np(1-p)}$$

KEY EXAMPLE

Of all colon cancers, 80% are cured if detected early. Among a group of 12 patients with new diagnoses, the mean and standard deviation for the number of cures are calculated:

$$\mu = np = 12(.8) = 9.6 \text{ and } \sigma = \sqrt{np(1-p)} = \sqrt{12(.8)(.2)} = 1.39$$

Key 25 Simulation

OVERVIEW *Simulation is a tool for estimating probabilities of complex events. There are many possible ways of setting up any given simulation; whatever the method, however, presenting a design in clear and unambiguous language is a primary goal.*

KEY EXAMPLE

If there are 30 people in a room, what is the probability that at least 2 have the same birthday?

Answer: Randomly pick 30 numbers between 1 and 365 (for example, on the TI-83 go to MATH $\rightarrow$ PRB $\rightarrow$ randInt and enter randInt(1,365,30)) and check if there are any duplicates. Repeat (simply hit ENTER again on the TI-83) this process many times and note the proportion of times that you find a duplicate.

KEY EXAMPLE

The female of some animal species must choose a mate from among a sequence of competing males. Suppose there are 10 males, and the female's strategy is to reject the first n males and then pick the first of the remaining males better than those rejected males. What should n be to maximize the probability of her picking the best male?

Answer: To test each n, for example, $n = 3$, proceed as follows. Put ten cards with the numbers 1 through 10 (where 10 is the highest) in a box. Randomly pick three cards to reject, and then pick cards, one at a time, stopping when a card is picked, larger than the three rejects (or until you're stuck with the last card, whatever it is). Repeat this experiment over and over, noting the proportion of times the female ends up with the number 10 card.

KEY EXAMPLE

Suppose the Cubs have a 40% chance of winning any given game against the Yankees. If these two teams meet in the World Series, what is the probability that the Cubs will win the Series—that is, that they will win four games before the Yankees do?

Answer: In a table of random numbers, let 0, 1, 2, and 3 represent Cub wins and 4, 5, 6, 7, 8, and 9 represent Yankee wins. Given any row in the table, read off the numbers one-by-one and note which comes first, four numbers from among {0,1,2,3} or four numbers from among {4,5,6,7,8,9}. Repeat this procedure many times, noting the proportion of times the Cubs win four games before the Yankees do.

KEY EXAMPLE

One floor of a large department store has aisles represented by a grid of 10 horizontal by 10 vertical lines. If a mother and lost child start at random locations, and each moves randomly among all available directions, how many "moves" will it take, on average, for the mother to spot the child (be on same horizontal or vertical line)?

Answer: Using the TI-83, randInt(1,10,2) gives the starting *x*- and *y*-coordinates of the mother and hitting ENTER a second time gives the starting location of the child. A "move" then consists of randomly picking a direction for each to move one unit, noting that randInt(1,4) and letting 1=north, 2=west, 3=south, and 4=east will give a direction to move from interior locations, whereas randInt(1,3) for edges and randInt(1,2) for corners are necessary. After each simultaneous move, check to see if the child is spotted. Note how many moves are made before the mother spots the child. Repeat this procedure many times and then determine the average of the results.

Key 26 Theme exercises with answers

OVERVIEW *Sample questions of the type that might appear on homework assignments and tests are presented with answers.*

- If the probability of a defective lightbulb is .1, what is the probability that exactly three out of eight lightbulbs are defective?

Answer:

$$C(8,3)(.1)^3(.9)^5 = \frac{8!}{3!5!}(.1)^3(.9)^5 = .0331$$

[Or on the TI-83, binompdf(8,.1,3) = .0331.]

- A grocery store manager notes that 35% of the customers buying a particular product will make use of a store coupon to receive a discount. If seven people purchase the product, what is the probability that fewer than four will use a coupon?

Answer: In this situation, "fewer than four" means zero or one or two or three.

$C(7,0)(.35)^0(.65)^7 + C(7,1)(.35)^1(.65)^6 + C(7,2)(.35)^2(.65)^5$
$\qquad + C(7,3)(.35)^3(.65)^4$
$\quad = (.65)^7 + 7(.35)(.65)^6 + 21(.35)^2(.65)^5 + 35(.35)^3(.65)^4$
$\quad = .800$

[Or binomcdf(7,.35,3) = .800.]

- If 6% of the major success stories with regard to playing the stock market are due to illegal insider information passing, in a group of seven successful investors, what is the probability that exactly one was achieved dishonestly? That at least one was achieved dishonestly?

Answer:

P(dishonest gain) = .06 so P(honest gain) = .94.

P(exactly one dishonest gain) = $7(.06)(.94)^6$ = .2897

P(at least one dishonest gain) = $1 - P$(all honest gains)
$\qquad\qquad\qquad\qquad\qquad = 1 - (.94)^7 = .3515$

[Or binompdf(7,.06,1) = .2897 and 1 − binompdf(7,.06,0) = .3515.]

- An ambulance service calculates the number of calls per day to be the following random variable:

Calls per day:	0	1	2	3	4	5	6	7	8
Probability:	.12	.15	.18	.26	.10	.08	.06	.03	.02

Calculate the expected value and variance for the calls per day variable.

Answer:

$$\mu = \Sigma x P(x)$$
$$= 0(.12) + 1(.15) + 2(.18) + 3(.26) + 4(.10) + 5(.08) + 6(.06) + 7(.03) + 8(.02)$$

$$= 2.82$$

$$\sigma^2 = \Sigma(x - \mu)^2 P(x) = \Sigma x^2 P(x) - \mu^2$$
$$= 0(.12) + 1(.15) + 4(.18) + 9(.26) + 16(.10) + 25(.08) + 36(.06) + 49(.03) + 64(.02) - (2.82)^2$$

$$= 3.7676$$

- Sixty percent of all new-car buyers choose automatic transmissions. For a group of five new car buyers, calculate the mean and standard deviation for the number of buyers choosing automatics.

Answer:

$$\mu = np = 5(.6) = 3.0$$

and

$$\sigma = \sqrt{np(1 - p)} = \sqrt{5(.6)(.4)} = 1.1$$

Note that these values could have been calculated in a more involved way:

$$\mu = \Sigma x P(x)$$
$$= 0[(.4)^5] + 1[5(.6)(.4)^4] + 2[10(.6)^2 (.4)^3] + 3[10(.6)^3 (.4)^2] + 4[5(.6)^4 (.4)] + 5[(.6)^5]$$
$$= 3.0$$

$$\sigma = \sqrt{\Sigma(x - \mu)^2 P(x)}$$
$$= \sqrt{9[.01024] + 4[.07680] + 1[.23040] + 0[.34560] + 1[.25920] + 4[.07776]}$$
$$= 1.1$$

Theme 4   PROBABILITY
DISTRIBUTIONS

*T*his theme presents several probability distributions of general interest in statistics. Such knowledge is necessary for future study and is also immediately useful as a decision-making tool for certain classes of problems.

For example, knowing the probability of finding oil, given certain geological conditions, we can use the *binomial distribution* to calculate the probability of various numbers of positive strikes for a given number of test sitings. Knowing the average number of Supreme Court vacancies during previous presidential terms, we can use the *Poisson distribution* to calculate the probability of various numbers of vacancies arising during the next 4-year term. Knowing the mean and variance of heights of U.S. Marines, we can use the *normal distribution* to calculate the probability that any Marine has a height greater than a specified value.

The binomial distribution arises from the concepts and calculations of preceding keys. The Poisson distribution can be viewed as a limiting case of the binomial when n is large and p is small. The normal distribution can be viewed as a limiting case of the binomial when p is constant but n increases without bound. Finally, it cannot be overstressed that, even if there were very few "naturally" occurring normal distributions, the normal has tremendous value in that it describes the distribution found in a wide range of statistical experiments, investigations, and studies. This aspect will be the focus of Themes 6 and 7.

INDIVIDUAL KEYS IN THIS THEME

Key 27 Binomial distributions

OVERVIEW *The concept of binomial probability from Key 20 can be expanded to consider tables, histograms, and the notions of mean and standard deviation.*

To review, a binomial experiment is characterized as follows: (1) the experiment consists of *n* identical trials; (2) each trial has the same two outcomes, commonly called *success* and *failure,* with probabilities *p* and *1 – p,* respectively, (3) the probabilities of success and failure remain the same from trial to trial; that is, the outcome of any one trial has no effect on the outcome of any other trial; (4) the mean is $\mu = np;$ and (5) the standard deviation is $\sigma = \sqrt{np(1-p)}$.

KEY EXAMPLE

The probability is .6 that a well driller will find water at a depth of less than 100 feet in a certain area. Wells are to be drilled for six new home owners. What is the complete probability distribution for the number of wells under 100 feet?

Answer: If $P(x)$ represents the probability of *x* wells under 100 feet, then

$$P(0) = \quad 1(.6)^0(.4)^6 = .004$$
$$P(1) = \quad 6(.6)^1(.4)^5 = .037$$
$$P(2) = \quad 15(.6)^2(.4)^4 = .138$$
$$P(3) = \quad 20(.6)^3(.4)^3 = .276$$
$$P(4) = \quad 15(.6)^4(.4)^2 = .311$$
$$P(5) = \quad 6(.6)^5(.4)^1 = .187$$
$$P(6) = \quad 1(.6)^6(.4)^0 = \underline{.047}$$
$$ 1.000$$

This probability distribution has mean $\mu = np = 6(.6) = 3.6$ and standard deviation $\sigma = \sqrt{np(1-p)} = \sqrt{6(.6)(.4)} = 1.2$. It is useful to label two horizontal axes, one with raw scores and one with *z*-scores. In this case,

z-scores of 0, 1, and 2 correspond to 3.6, 4.8, and 6.0, while z-scores of −1, −2, and −3 correspond to 2.4, 1.2, and 0, respectively.

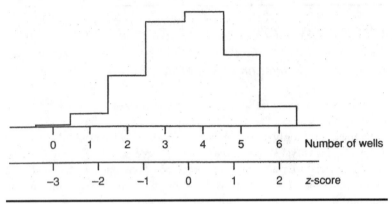

A common decision-making problem arising in manufacturing situations involves whether or not to accept a shipment of raw materials or of finished products. Concerns over quality control are sometimes handled by inspecting a sample before allowing a whole shipment to proceed. An operational rule is chosen through which the shipment will be accepted only if fewer than some specified number of defects are found in the sample. Deciding on sample size and allowable number of defects is specific to the given situation and must take into consideration time, money, quality needs, and so on. To analyze a proposed sampling plan it is important to calculate the probability of shipment acceptance, given various possible shipment defect levels.

KEY EXAMPLE

Suppose the decision-making rule is to pick a sample of size $n = 10$, and accept the whole shipment if the sample contains at most one defective item. What is the probability of acceptance if the defect level of the shipment is actually 5%? 10%? 20%? 30%? 40%? 50%?

Answer: For $p = .05$, $P(\text{accept}) = P(0 \text{ def}) + P(1 \text{ def}) = C(10,0)(.05)^0(.95)^{10} + C(10,1)(.05)^1(.95)^9 = (.95)^{10} + 10(.05)(.95)^9 = .914$. Then:

For $p = .10$, $P(\text{accept}) = (.90)^{10} + 10(.10)(.90)^9 = .736$

For $p = .20$, $P(\text{accept}) = (.80)^{10} + 10(.20)(.80)^9 = .376$

For $p = .30$, $P(\text{accept}) = (.70)^{10} + 10(.30)(.70)^9 = .149$

For $p = .40$, $P(\text{accept}) = (.60)^{10} + 10(.40)(.60)^9 = .046$

For $p = .50$, $P(\text{accept}) = (.50)^{10} + 10(.50)(.50)^9 = 0.11$

The graph of the above values is called the *operating characteristic curve* for this sampling plan.

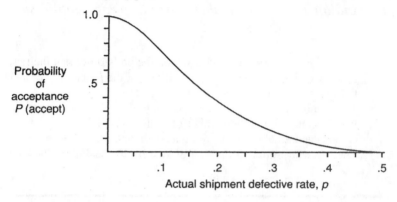

Note that, if $p = 0$, that is, there are no defective items in the entire shipment, then $P(\text{accept}) = 1$, while if $p = 1.0$, that is, the entire shipment is defective, then $P(\text{accept}) = 0$.

Key 28 Geometric distributions

OVERVIEW *In a binomial experiment, the **geometric distribution** shows the number of trials needed until a success is achieved.*

- If p is the probability of success, then the probability that the first success will appear on the k-th trial, that is, the probability of $k-1$ failures followed by a success, is $(1-p)^{k-1} p$.
- The conditions for a geometric variable differs from a binomial variable only in that the number of trials is not fixed.
- Geometric probabilities decrease as the number of trials increase. The greatest probability is that of success happening on the first trial!

KEY EXAMPLE

Suppose the probability that a company will land a contract is .3. What is the probability that it will first land a contract on the third bid? No later than the third bid?

Answer: P(first on third bid) $= (.7)^2(.3) = .147$ and P(third or sooner) $= .3 + (.7)(.3) + (.7)^2(.3) = .657$. [On the TI-83, geometpdf(.3,3) = .147 and geometcdf(.3,3) = .657.]

Key 29 Poisson distributions

OVERVIEW   *The limiting case of the binomial when* n *is large and* p *is small is called the **Poisson distribution**. To calculate a Poisson probability, one needs to know only the average,* μ.

A probability distribution given by $P(x \text{ successes}) = \dfrac{\mu^x}{x!}e^{-\mu}$ where μ is the mean and $e = 2.71828\ldots$, is called a **Poisson probability distribution**.

KEY EXAMPLE

A grocery store manager knows that his store sells an average of three cans of artichoke hearts per week. Assuming that this situation is described by the formula given above, what is the probability that no can of artichoke hearts is sold in 1 week? The probability of one can in a week? Two cans in a week? One can in half a week? Three cans in 2 weeks?

Answer:

> For 1 week the average is three cans, so
>> $P(\text{no can in 1 week})\ \ = e^{-3}\ \ \ \ \ \ \ = .050$
>> $P(\text{one can in 1 week})\ = 3e^{-3}\ \ \ \ \ = .149$
>> $P(\text{two cans in 1 week}) = (3^2/2)e^{-3} = .224$
>
> For a half week the average is $0.5(3) = 1.5$, so
>> $P(\text{one can in 0.5 week}) = 1.5e^{-1.5} = .335$
>
> For 2 weeks the average is $2(3) = 6$, so
>> $P(\text{three cans in 2 weeks}) = (6^3/3!)e^{-6} = .089$

In many examples the probabilities are multiplied by some total number to derive numerical distributions.

KEY EXAMPLE

One of the first noted examples of a Poisson distribution concerns the number of deaths from horse kicks in various corps of the German army in the late 1800s. More specifically, during the 20-year period 1875 to

1894, among the 14 cavalry corps of the German army, there was an average 0.7 deaths per corps per year as the result of horse kicks. If the deaths followed a Poisson distribution, then the descriptive probabilities per corps per year would be:

$$P(0 \text{ deaths}) = e^{-0.7} \qquad\qquad = .497$$
$$P(1 \text{ death}) = 0.7e^{-0.7} \qquad\quad = .348$$
$$P(2 \text{ deaths}) = [(0.7)^2/2]e^{-0.7} = .122$$
$$P(3 \text{ deaths}) = [(0.7)^3/3!]e^{-0.7} = .028$$
$$P(4 \text{ deaths}) = [(0.7)^4/4!]e^{-0.7} = \underline{.005}$$
$$1.000$$

If these are the correct probabilities, then the expected numerical distribution, found by multiplying the probabilities by 280 (20 years × 14 corps = 280 groups), is as follows:

Deaths	Predicted number of groups
0	.497 × 280 = 139.2
1	.348 × 280 = 97.4
2	.122 × 280 = 34.2
3	.028 × 280 = 7.8
4	.005 × 280 = 1.4
	280.0

How accurate is this description? Here are the actual figures:

Deaths	Actual number of groups
0	144
1	91
2	32
3	11
4	2

To calculate a binomial distribution, we must be given n, the number of cases under consideration. In other words, to find the probability of a certain number of successes, we must know the corresponding number of failures. However, there are situations in which the number of failures is impossible to determine. For example, in determining the probability that a hospital will have two appendectomy cases in one day, it makes little sense to ask how many appendicitis cases will not occur. In calculating the probability that a baseball team will score five runs in a game, it is meaningless to ask how many runs will not be scored. And in finding the probability that there will be ten incoming telephone calls at a switchboard, it is impossible to say how many calls will not come in. Fortunately there are many cases in which n is very large, p is very small, and the mean or average, $\mu = pn$ is both moderate and known. Then the Poisson distribution can be applied as above.

KEY EXAMPLE

A large number n of planes fly into a major airport, and the probability p that any particular plane is on the runway at any particular moment in time is small. However, the average number np of planes waiting in line on the runway at noon each day may be determined without evaluating either p or n. Once $\mu = np$ is calculated, then the Poisson formula may be applied.

The Poisson is used in certain types of hypothesis tests.

KEY EXAMPLE

Observations taken before a company began to dump pollutants into a river indicated that an average of three trout per hour swim past the dump site. An inspector plans to spend an hour at the site, and will issue a warning if she sees fewer than three trout. If the pollutants have no effect on the trout, and so the mean per hour is still three, what is the probability that there will be a warning?

Answer:

$$P(\text{less than } 3) = P(0) + P(1) + P(2) = e^{-3} + 3e^{-3} + [3^2/2]e^{-3} = .423$$

If the pollutants really kill half of the trout, what is the probability that a warning does or does not result?

Answer: In this case, the mean is $.5(3) = 1.5$, and thus we have

$$P(\text{warning}) = e^{-1.5} + 1.5e^{-1.5} + [(1.5)^2/2]e^{-1.5} = .809$$

$$P(\text{no warning}) = 1 - .809 = .191$$

Note that with this inspection procedure, there still might be a warning (.423 probability) even if the pollutant is harmless, and there might be no warning (.191 probability) even if the pollutant kills half the fish.

Key 30 Normal distributions

OVERVIEW *The normal curve is bell shaped and symmetrical with an infinite base. There are long, flattened tails that cover many values but only a small proportion of the area. The mean here is the same as the median and is located at the center. It is convenient to measure distances under the normal curve in terms of z-scores (fractions or multiples of standard deviations from the mean).*

Table A in the Appendix gives proportionate areas under the normal curve. Because of the symmetry of the curve, it is sufficient to give areas from the mean to positive z values. Table A shows, for example, that between the mean and a z-score of 1.2 there is .3849 of the area:

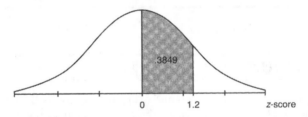

while .4834 of the area is between the mean and a z-score of –2.13.

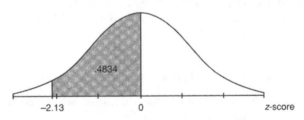

[On the TI-83, normalcdf(0,1.2) = .3849 and normalcdf(–2.13,0) = .4834.]

KEY EXAMPLE

The life expectancy of a particular brand of lightbulbs is normally distributed with a mean of 1500 hours and a standard deviation of 75 hours.

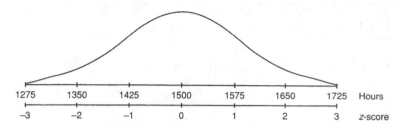

What is the probability that a bulb will last between 1500 and 1650 hours?

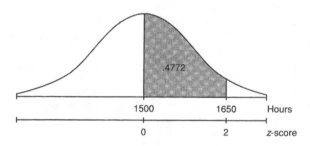

Answer: The z-score of 1500 is 0, the z-score of 1650 is $(1650 - 1500)/75 = 2$, and 2 in Table A gives a probability of .4772. [normalcdf(0,2) = .4772 and also normalcdf(1500,1650,1500,75) = .4772.]

What percentage of the lightbulbs will last between 1485 and 1500 hours?

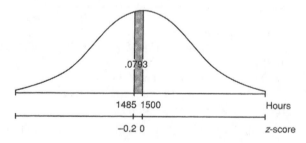

Answer: The z-score of 1485 is $(1485 - 1500)/75 = -0.2$, and 0.2 in Table A gives a probability of .0793 or 7.93%. [normalcdf(-0.2,0) = .0793 and also normalcdf(1485,1500,1500,75) = .0793.]

What is the probability that a bulb will last between 1416 and 1677 hours?

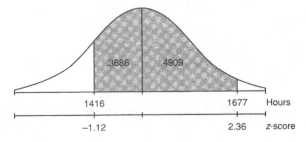

Answer: The *z*-score of 1416 is (1416 − 1500)/75 = −1.12, and the *z*-score of 1677 is (1677 − 1500)/75 = 2.36. In Table A, 1.12 gives a probability of .3686, and 2.36 gives a probability of .4909. The total probability is .3686 + .4909 = .8595. [normalcdf(−1.12,2.36) = .8595 and also normalcdf(1416,1677,1500,75) = .8595.]

What is the probability that a lightbulb will last between 1563 and 1648 hours?

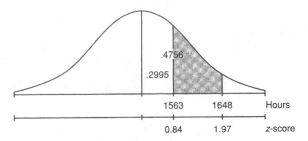

Answer: The *z*-score of 1563 is (1563 − 1500)/75 = 0.84, and, the *z*-score of 1648 is (1648 − 1500)/75 = 1.97. In Table A, 0.84 and 1.97 give probabilities of .2995 and .4756. Between 1563 and 1648 there is a probability of .4756 − .2995 = .1761. [normalcdf(0.84,1.97) = .1760 and also normalcdf(1563,1648,1500,75) = .1762.]

What is the probability that a lightbulb will last less than 1410 hours?

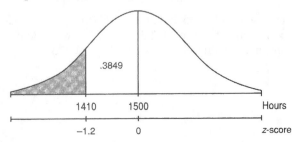

Answer: The z-score of 1410 is (1410 − 1500)/75 = −1.2. In Table A, 1.2 gives a probability of .3849. The probability of being less than 1500 is .5, and so the probability of being less than 1410 is .5 − .3849 = .1151. [normalcdf(−100,−1.2) = .1151 and also normalcdf(0,1410,1500,75) = .1151.]

KEY EXAMPLE

A packing machine is set to fill a cardboard box with a mean average of 16.1 ounces of cereal. Suppose the amounts per box form a normal distribution with standard deviation equal to 0.04 ounce.

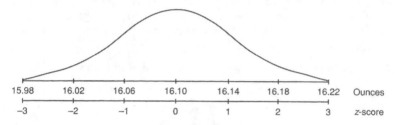

What percent of the boxes end up with at least 1 pound of cereal?

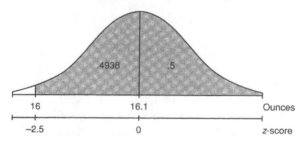

Answer: The z-score of 16 is (16 − 16.1)/0.04 = −2.5, and 2.5 in Table A gives a probability of .4938. The probability of more than 1 pound is .4938 + .5 = .9938 or 99.38%. [normalcdf(−2.5,100) = .9938 and also normalcdf(16,1000,16.1,0.04) = .9938.]

Ten percent of the boxes will contain above what number of ounces?

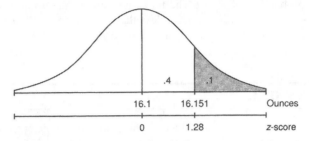

Answer: In Table A, we note that .4 area (actually .3977) is found between the mean and a *z*-score of 1.28, so to the right of a 1.28 *z*-score must be 10% of the area. Converting the *z*-score of 1.28 into a raw score yields 16.1 + 1.28(0.04) = 16.151 ounces. [invNorm(.9) = 1.2816 and also invNorm(.9,16.1,0.04) = 16.151.]

Eighty percent of the boxes will contain above what number of ounces?

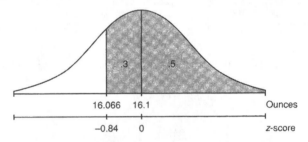

Answer: In Table A, we note that .3 area (actually .2995) is found between the mean and a *z*-score of 0.84, so to the right of a –0.84 *z*-score must be 80% of the area. Converting the *z*-score of –0.84 into a raw score yields 16.1 – 0.84(0.04) = 16.066 ounces. [invNorm(.2) = –0.8416 and also invNorm(.2,16.1,0.04) = 16.066.]

Key 31 Commonly used probabilities
and their *z*-scores

OVERVIEW *There is often an interest in the limits enclosing some specified middle percentage of the data and in values with particular percentile rankings.*

For future reference, we note, in terms of *z*-scores, the limits most often asked for.

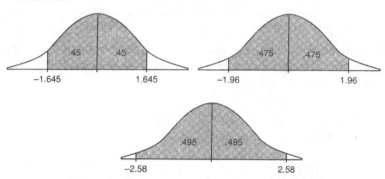

Since .45 + .45 = .90, then 90% of the values are between *z*-scores of −1.645 and +1.645. Since .475 + .475 = .95, then 95% of the values are between *z*-scores of −1.96 and +1.96. Since .495 + .495 = .99, then 99% of the values are between *z*-scores of −2.58 and +2.58.

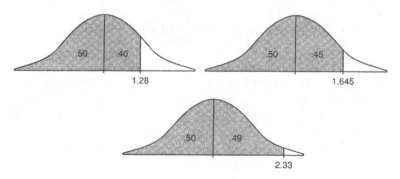

Thus, 90% of the values are below a z-score of 1.28, 95% of the values are below a z-score of 1.645, and 99% of the z-scores are below a z-score of 2.33.

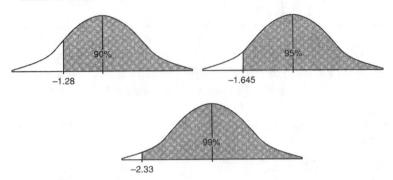

Similarly, 90% of the values are above a z-score of −1.28, 95% of the values are above a z-score of −1.645, and 99% of the values are above a z-score of −2.33.

It is also useful to note the percentages corresponding to values falling between integer z-scores.

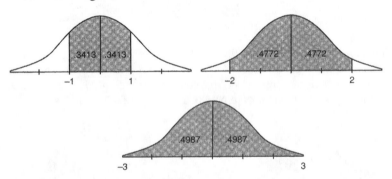

For example, .3414 + .3413 = .6826, and so 68.26% of the values are between z-scores of −1 and +1. Also, .4772 + .4772 = .9544, and so 95.44% of the values are between z-scores of −2 and +2. And .4987 + .4987 = .9974, and so 99.74% of the values are between z-scores of −3 and +3.

KEY EXAMPLE

Suppose that the average height of adult males in a particular locality is 70 inches with a standard deviation of 2.5 inches. If the distribution is normal, then the middle 95% of males are between what two heights?

Answer: As noted above, the critical z-scores are ±1.96, so the two limiting heights are 1.96 standard deviations from the mean. Therefore, 70 ± 1.96(2.5) = 70 ± 4.9 or from 65.1 to 74.9 inches. [invNorm(.025,70,2.5) = 65.1 and invNorm(97.5,70,2.5) = 74.9.]

Ninety percent of the heights are below what value?

Answer: The critical z-score is 1.28, and so the height in question is 70 + 1.28(2.5) = 70 + 3.2 = 73.2 inches. [invNorm(.9,70,2.5) = 73.2.]

Ninety-nine percent of the heights are above what value?

Answer: The critical z-score is –2.33, so the height in question is 70 – 2.33(2.5) = 70 – 5.825 = 64.175 inches. [invNorm(.01,70,2.5) = 64.184.]

What percentage of the heights are between z-scores of ±1? Of ±2? Of ±3?

Answer: 68.26%, 95.44%, and 99.74%, respectively. [normalcdf(–1,1) = .6827, normalcdf(–2,2) = .9545, and normalcdf(–3,3) = .9973.]

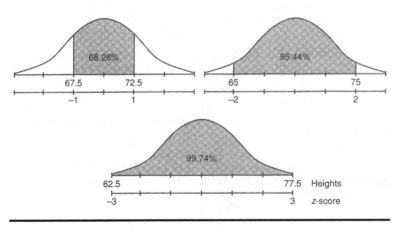

Key 32 Finding means and standard deviations in normal distributions

OVERVIEW *Knowing that a distribution is normal allows for calculations of the mean μ and the standard deviation σ, using percentage information from the population.*

KEY EXAMPLE

Given a normal distribution with a mean of 25, what is the standard deviation if 18% of the values are above 29?

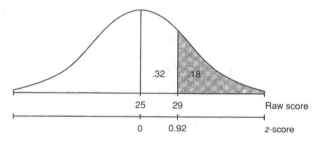

Answer: Looking for a .32 probability in Table A, we note that the corresponding *z*-score is 0.92. Thus 29 – 25 = 4 is equal to 0.92 standard deviation, that is, 0.92σ = 4, and σ = 4/0.92 = 4.35.

KEY EXAMPLE

Given a normal distribution with a standard deviation of 10, what is the mean if 21% of the values are below 50?

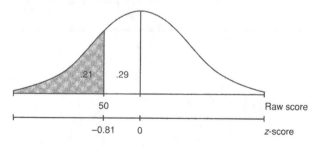

Answer: Looking for a .29 probability in Table A leads to a *z*-score of –0.81. Thus 50 is –0.81 standard deviation from the mean, and so μ = 50 + 0.81(10) = 58.1.

KEY EXAMPLE

Given a normal distribution with 80% of the values above 125 and 90% of the values above 110, what are the mean and the standard deviation?

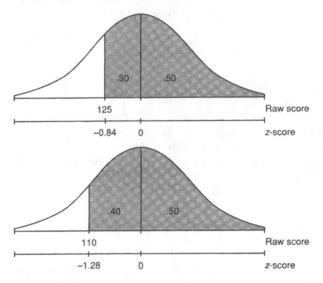

Answer: Table A gives critical *z*-scores of –0.84 and –1.28. Thus we have $(125 – μ)/σ = –0.84$ and $(110 – μ)/σ = –1.28$. Solving the system $\{125 – μ = –0.84σ, 110 – μ = –1.28σ\}$ simultaneously gives μ = 153.64 and σ = 34.09.

Key 33 Normal approximation to the binomial

OVERVIEW *Many practical applications of the binomial involve examples where* n *is large. However, for large* n, *binomial probabilities can be quite messy to calculate. Since the normal can be viewed as a limiting case of the binomial, it is natural to use the normal to approximate the binomial in appropriate situations.*

The binomial takes values only at integers, while the normal is continuous with probabilities corresponding to areas over intervals. Therefore, we must set down some technique for converting from one distribution to the other. For approximation purposes we do as follows. Each binomial probability will correspond to the normal probability over a unit interval centered at the desired value. Thus, for example, to approximate the binomial probability of eight successes we determine the normal probability of being between 7.5 and 8.5.

KEY EXAMPLE

Suppose that 15% of the cars coming out of an assembly plant have some defect. In a delivery of 40 cars what is the probability that exactly 5 cars have defects?

Answer: The actual answer is $(40!/35!5!)(.15)^5(.85)^{35}$ but as can be seen this involves a nontrivial calculation. To approximate the answer using the normal, we first calculate the mean μ and the standard deviation σ as follows:

$$\mu = np = 40(.15) = 6$$
$$\sigma = \sqrt{np(1-p)} = \sqrt{40(.15)(.85)} = 2.258$$

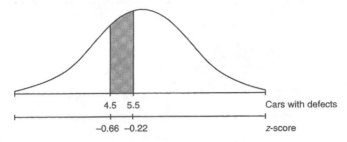

We then calculate the appropriate z-scores: $(4.5 - 6)/2.258 = -0.66$ and $(5.5 - 6)/2.258 = -0.22$. Looking up the corresponding probabilities in Table A gives a final answer of $.2454 - .0871 = .1583$. (The actual answer is $.1692$.)

Even more useful are approximations relating to probabilities over intervals.

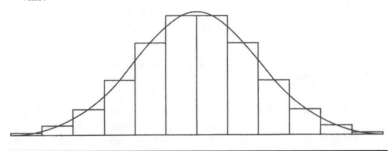

KEY EXAMPLE

If 60% of the population support massive federal budget cuts, what is the probability that in a survey of 250 people at most 155 people support the cuts?

Answer: The actual answer is the sum of 156 binomial expressions:

$$(.4)^{250} + \cdots + \frac{250!}{155!95!}(.6)^{155}(.4)^{95}$$

However, a good approximation can be obtained quickly and easily using the normal. We calculate μ and σ:

$$\mu = np = 250(.6) = 150$$
$$\sigma = \sqrt{np(1-p)} = \sqrt{250(.6)(.4)} = 7.746$$

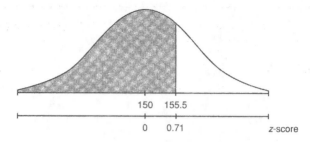

The binomial of at most 155 successes corresponds to the normal probability of ≤155.5. The z-score of 155.5 is $(155.5 - 150)/7.746 = 0.71$, and using Table A leads to a final answer of $.5000 + .2611 = .7611$.

Is the normal a good approximation? The answer, of course, depends on the error tolerances in particular situations. However, a general rule of thumb is that the normal is a "good" approximation to the binomial whenever np and $n(1 - p)$ are both greater than 5 (some statisticians use 10).

KEY EXAMPLE

A form of cancer is fatal in 30% of all diagnosed cases. A new drug is tried on 200 patients with this disease, and researchers will judge the new medication effective if at least 150 of the patients recover. If the medication has no effect, what is the probability that at least 150 patients recover?

Answer: The actual answer is the sum of 51 binomial expressions:

$$\frac{200!}{150!50!}(.7)^{150}(.3)^{50} + \frac{200!}{151!49!}(.7)^{151}(.3)^{49} + \cdots + (.7)^{200}$$

That would be very tedious to calculate. However, an approximate answer using the normal is readily calculated.

$$\mu = np - 200(.7) = 140$$
$$\sigma = \sqrt{np(1 - p)} = \sqrt{200(.7)(.3)} = 6.48$$

The binomial probability of 150 successes corresponds to the normal probability on the interval from 149.5 to 150.5, and thus the binomial probability of at least 150 successes corresponds to the normal probability of ≥ 149.5. The z-score of 149.5 is $(149.5 - 140)/6.48 = 1.47$. Using the normal probability table gives a final answer of $5.000 - .4292 = .0708$.

Key 34 Theme exercises with answers

OVERVIEW *Sample questions of the type that might appear on homework assignments and tests are presented with answers.*

- A baseball player, with a batting average of .250, has 12 official at-bats in a three-game series. What is the probability distribution for the number of hits he makes? Display in a histogram.

  Answer: The probability of a hit is .250, and we have the following:

  $P(0) = C(12,0)(.250)^0(.750)^{12} = 1(.75)^{12}$ $= .032$
  $P(1) = C(12,1)(.250)^1(.750)^{11} = 12(.25)(.75)^{11}$ $= .127$
  $P(2) = C(12,2)(.250)^2(.750)^{10} = 66(.25)^2(.75)^{10} = .232$
  $P(3) = C(12,3)(.250)^3(.750)^9 = 220(.25)^3(.75)^9 = .258$
  $P(4) = C(12,4)(.250)^4(.750)^8 = 495(.25)^4(.75)^8 = .194$
  $P(5) = C(12,5)(.250)^5(.750)^7 = 792(.25)^5(.75)^7 = .103$
  $P(6) = C(12,6)(.250)^6(.750)^6 = 924(.25)^6(.75)^6 = .040$
  $P(7) = C(12,7)(.250)^7(.750)^5 = 792(.25)^7(.75)^5 = .012$
  $P(8) = C(12,8)(.250)^8(.750)^4 = 495(.25)^8(.75)^4 = .002$

  To three decimal places, $P(9) = P(10) = P(11) = P(12) = 0$.

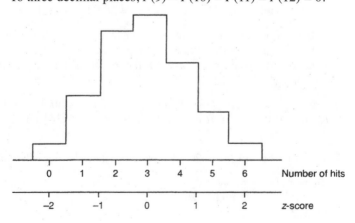

  The mean number of hits is
  $$\mu = np = 12(.25) = 3$$

and the standard deviation is

$$\sigma = \sqrt{np(1-p)} = \sqrt{12(.25)(.75)} = 1.5$$

Thus, z-scores of 0, 1, and 2 correspond to 3, 4.5, and 6, while z-scores of -1 and -2 correspond to 1.5 and 0, respectively.

- In a large northeastern town the average number of business bankruptcies per week is 1.5. What is the probability of no bankruptcy in a week? Of one bankruptcy? Of two bankruptcies? Of three?

Answer: There are a large number of businesses, the probability of bankruptcy is small, and the average is known; therefore, we apply the Poisson distribution to obtain:

$$P(\text{no bankruptcy}) = e^{-1.5} = .223$$
$$P(1 \text{ bankruptcy}) = 1.5e^{-1.5} = .335$$
$$P(2 \text{ bankruptcies}) = [(1.5)^2/2]e^{-1.5} = .251$$
$$P(3 \text{ bankruptcies}) = [(1.5)^3/3!]e^{-1.5} = .126$$

What is the probability of no bankruptcy in a 2-week period?

Answer: The average for 2 weeks is 2(1.5) = 3, so

$$P(0 \text{ bankruptcy in 2 weeks}) = e^{-3} = .0498$$

What is the probability of four bankruptcies in a 3-week period?

Answer: The average for 3 weeks is 3(1.5) = 4.5, so

$$P(4 \text{ bankruptcies in 3 weeks}) = [(4.5)^4/4!]e^{-4.5} = .1898$$

- The resistors in a shipment have an average resistance of 200 ohms with a standard deviation of 5 ohms. If a normal distribution is assumed, what percentage of the resistors have resistances between 195 and 212 ohms?

Answer: The z-scores of 195 and 212 are (195 − 200)/5 = −1 and (212 − 200)/5 = 2.4, respectively. Table A gives .3413 + .4918 = .8331 = 83.31%.

Thirty percent of the resistors have resistances below what value?

Answer: For a probability of .5 − .3 = .2, Table A gives a z-score of −0.52. Converting to a raw score gives 200 − 0.52(5) = 197.4 ohms.

The middle 95% have resistances between what two values?

Answer: The critical z-scores ±1.96 convert to 200 ± 1.96(5) = 200 ± 9.8 or between 190.2 and 209.8 ohms.

- Airline companies know that 4% of all reservations received will be no-shows, so they overbook accordingly. Suppose that there are 126 seats on a plane, and the airline books 130 reservations. What is the probability that more than 126 confirmed passengers will show up? In other words, what is the probability that the number of no-shows will be 3 or less? Solve using normal approximation to the binomial.

Answer: Both $np = 130(.04) = 5.2$ and $n(1-p) = 130(.96) = 124.8$ are greater than 5, so it is reasonable to determine a normal approximation.

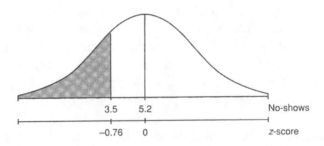

The normal approximation is calculated using $n = 130$ and $p = .04$ to derive $\mu = 5.2$ and $\sigma = 2.2343$. The z-score of 3.5 is $(3.5 - 5.2)/2.2343 = -0.76$, and the probability approximation is $.5 - .2764 = .2236$.

The actual answer is the sum of four binomial probabilities:

$$\frac{130!}{127!3!}(.04)^3(.96)^{127} + \frac{130!}{128!2!}(.04)^2(.96)^{128} + 130(.04)(.96)^{129}$$
$$+(.96)^{130} = .2323$$

Theme 5   PLANNING A STUDY

*I*n the real world, time and cost considerations usually make it impossible to analyze an entire population. In studying statistics we learn how to estimate *population* characteristics by considering *samples*. We want to be confident that samples we choose represent the population fairly. Analyzing the data with computers is usually easier than gathering the data, but the frequently quoted "Garbage in, garbage out" applies here. Nothing can help if the data are badly collected. Unfortunately, many of the statistics with which we are bombarded by newspapers, radio, and television are based on poorly designed data-collection procedures.

A *census* is a complete enumeration of an entire population. However, a well-designed, well-conducted *sample survey* is far superior to a poorly designed study involving a complete census. In this theme we look at a variety of sampling methods and sources of bias in surveys. Then we compare *observational studies with experiments,* and list the properties of a good experiment.

INDIVIDUAL KEYS IN THIS THEME

Key 35 Sampling methods

OVERVIEW   *Most data collection involves observation-al studies, not controlled experiments. Furthermore, although most data collection has some purpose, many studies come to mind after the data have been assembled and examined. For data collection to be useful, the resulting sample must be representative of the population under consideration. Standard sampling methods include* **simple random sampling, systematic sampling, stratified sampling, proportional sampling,** *and* **multistage sampling.**

How can a representative sample be chosen? One technique is to write the name of each member of the population on a card, mix the cards thoroughly in a large box, and pull out a specified number of cards. Unfortunately, this method is usually too time consuming and too costly, and bias might still creep in if the mixing is not thorough. A *simple random sample* (SRS), that is, **one in which every possible sample of the desired size has an equal chance of being selected,** can more easily be obtained by assigning a number to everyone in the population and using a random number table or having a computer generate random numbers to indicate choices.

A simple random sample may be the ideal, but time- and cost-saving modifications are often called for, which leads to other standard sampling procedures. *Systematic sampling* involves randomly listing the population in order and then picking every tenth, hundredth, or thousandth, and so on, person from the list. In *stratified sampling* the population is divided into representative groups called *strata,* and random samples of people from all strata are chosen. In *proportional sampling* the sizes of the random samples from each stratum depend on the proportion of the total population represented by the stratum. *Multistage sampling* involves dividing the population into groupings, subdividing each grouping, selecting a random sampling of the subdivisions, and finally picking a random sample of people from the selected subdivisions. What is crucial in all these techniques is the presence of a methodical procedure that involves the use of chance and leaves no freedom of choice to the interviewers.

KEY EXAMPLE

Suppose a sample of 100 high school students from a school of size 5000 is to be chosen to determine their views on reinstating the draft. How might this sample be chosen? One method would be to have each student write his or her name on a slip of paper, put the papers in a box, and have the principal reach in and pull out 100 of the papers. However, questions could arise about how well the papers are mixed in the box. A method of arriving at a *simple random sample* would be to assign each student a number from 0001 to 5000 and then use a random number table, picking out four digits at a time and tossing out repeats and numbers over 5000.

Alternative procedures are as follows. From a list of the students, the surveyor could simply note every fiftieth name *(systematic sampling)*. Because students in each class have certain characteristics in common, the surveyor could use a random selection method to pick 25 students from each of the separate lists of freshmen, sophomores, juniors, and seniors *(stratified sampling)*. The researcher could separate the homerooms by classes; then randomly pick five freshmen homerooms, five sophomore homerooms, five junior homerooms, and five senior homerooms; and then randomly pick five students from each of the homerooms *(multistage sampling)*. The surveyor could separately pick random samples of males and females, the size of each of the two samples chosen according to the proportion of male and female students attending the school *(proportional sampling)*.

It should be noted that none of the alternative procedures results in a *simple random sample,* because every possible sample of size 100 does not have an equal chance of being selected.

Key 36 Sources of bias in surveys

OVERVIEW *Poorly designed sampling techniques result in bias—that is, in a tendency to favor the selection of certain members of a population. The way data are obtained is crucial—a large sample size cannot make up for a poor survey design or faulty collection techniques.*

- An often-cited example of *selection bias* is the *Literary Digest* opinion poll that predicted a landslide victory for Alfred Landon over Franklin D. Roosevelt in the 1936 presidential election. The *Digest* surveyed people with cars and telephones, but in 1936 only the wealthy minority, who mainly voted Republican, had cars and telephones.
- Examples of *nonresponse bias* are present in most mailed questionnaires. They tend to have very low response percentages, and it is often unclear which part of the population is responding.
- *Unintentional bias* often creeps in when the surveyor tries to systematically pick people representative of the whole population. For example, in the 1948 presidential election, the *Chicago Tribune* incorrectly called Thomas E. Dewey the winner over Harry S. Truman. Here the mistake was partly caused by misleading polls based on *quota sampling* that left the interviewers too much free choice in picking people to fill their quotas.
- *Voluntary response samples,* based on individuals who offer to participate, typically give too much emphasis to people with strong opinions. For example, radio call-in programs about controversial topics such as gun control, abortion, and school segregation do not produce meaningful data on what proportion of the population favor or oppose related issues.
- *Convenience samples,* based on choosing individuals who are easy to reach, are also suspect. For example, interviews at shopping malls tend to produce data highly unrepresentative of the entire population.
- The wording of questions or the very questions themselves can lead to *response bias*. People often don't want to be perceived as having unpopular, unsavory, or illegal views and so may respond untruthfully when face-to-face with an interviewer or when filling out a questionnaire that is not anonymous.
- Sometimes people chosen for a survey simply refuse to respond or are unreachable or too difficult to contact. These situations lead to *nonresponse bias.*

- At other times certain people are left out of consideration, which results in *undercoverage bias*. For example, telephone surveys simply ignore all those possible subjects who don't have telephones, and door-to-door surveys ignore the homeless.

KEY EXAMPLE

Ann Landers, who wrote a daily advice column appearing in newspapers across the country, once asked her readers, "If you had it to do over again, would you have children?" Of the more than 10,000 readers who responded, 70% said no. What did this show?

Answer: This survey was meaningless because of voluntary response bias, which often overrepresents negative opinions. The people who chose to respond were most likely parents who were unhappy, so there was very little chance that the 10,000 respondents were representative of the population.

KEY EXAMPLE

Two possible wordings for a questionnaire on gun control are as follows:

I. The United States has the highest rate of murder by handguns among all countries. Most of these murders are known to be crimes of passion or crimes provoked by anger between acquaintances. Are you in favor of a seven-day cooling-off period between the filing of an application to purchase a handgun and the resulting sale?

II. The United States has one of the highest violent crime rates among all countries. Many people want to keep handguns in their homes for self-protection. Fortunately, U.S. citizens are guaranteed the right to bear arms by the Constitution. Are you in favor of a seven-day waiting period between the filing of an application to purchase a needed handgun and the resulting sale?

The wording of these questions is clearly not neutral and will lead to response bias. The neutral way of asking this question would simply have been as follows: Are you in favor of a seven-day waiting period between the filing of an application to purchase a handgun and the resulting sale?

X   X
STORE: 0690   REG: 05/61   TRAN#: 1765
SALE   09/01/2008   EMP: 00355
****

price offered for the item during the 6 month period prior to the return will be refunded via a gift card.

Opened videos, music discs, cassettes, electronics, and audio books may only be exchanged for a replacement of the original item.

Periodicals, newspapers, out-of-print, collectible, pre-owned items, and gift cards may not be returned.

Returned merchandise must be in saleable condition.

BORDERS.

Returns to Borders Stores

Merchandise presented for return, including sale or marked-down items, must be accompanied by the original Borders store receipt or a Borders Gift Receipt. Returns must be completed within 30 days of purchase. For returns accompanied by a Borders Store Receipt, the purchase price will be refunded in the medium of purchase (cash, credit card or gift card). Items purchased by check may be returned for cash after 10 business days. For returns within 30 days of purchase accompanied by a Borders Gift Receipt, the purchase price (after applicable discounts) will be refunded via a gift card.

Merchandise unaccompanied by the original Borders store receipt, Borders Gift Receipt, or presented for return beyond 30 days from date of purchase, must be carried by Borders at the time of the return. The lowest price offered for the item during the 6 month period prior to the return will be refunded via a gift card.

Opened videos, music discs, cassettes, electronics, and audio books may only be exchanged for a replacement of the original item.

Periodicals, newspapers, out-of-print, collectible, pre-owned items, and gift cards may not be returned.

Returned merchandise must be in saleable condition.

BORDERS.

Returns to Borders Stores

Merchandise presented for return, including sale or marked-down items, must be accompanied by the original Borders store receipt or a Borders Gift Receipt.

No matter how well designed and well conducted a survey is, it still gives a sample *statistic* as an estimate for a population *parameter*. Different samples give different sample statistics, all of which are estimates for the same population parameter, and so error, called *sampling error,* is naturally present.

Key 37 Experiments versus observational studies

OVERVIEW *In an experiment we impose some change or treatment and measure the result or response. In an observational study we simply observe and measure something that has taken place or is taking place, while trying not to cause any changes by our presence. A sample survey is an observational study in which we draw conclusions about an entire population by considering an appropriately chosen sample to look at. An experiment often suggests a causal relationship, whereas an observational study may show only the existence of association.*

KEY EXAMPLE

A study is to be designed to determine whether daily calcium supplements benefit women by increasing bone mass. How can an observational study be performed? An experiment? Which is more appropriate here?

Answer: An observational study might interview and run tests on all women visiting a doctor's office during a certain time period. The bone mass measurements of those taking calcium supplements could then be compared with the measurements of those who say they don't take supplements.

An experiment could be performed by selecting some number of subjects, using chance to pick half to receive calcium supplements while the other half receives similar-looking *placebos,* and noting the difference in bone mass before and after treatment for each group.

The experimental approach is more appropriate here. With the observational study there could be many explanations for any bone mass difference noted between patients who take calcium and those who don't. For example, women who have voluntarily been taking calcium supplements might be precisely those who take better care of themselves in general and thus have higher bone mass for other reasons. The experiment tries to control for *lurking variables* by randomly giving half the subjects calcium.

KEY EXAMPLE

A study is to be designed to examine the life expectancies of tall people versus those of short people. Which is more appropriate, an observational study or an experiment?

Answer: An observational study, examining medical records of heights and ages at time of death, seems straightforward. An experiment where subjects are randomly chosen to be made short or tall, followed by recording age at death, would be groundbreaking (and, of course, nonsensical).

KEY EXAMPLE

A study is to be designed to examine the GPAs of college students who use marijuana regularly and those who don't. Which is more appropriate, an observational study or an experiment?

Answer: As much as some researchers might want to randomly require half the subjects to use an illegal drug, this would be unethical. The proper procedure here is an observational study, having students anonymously fill out questionnaires asking about marijuana usage and GPA.

Key 38 Planning and conducting
experiments

OVERVIEW *There are several primary principles deal-ing with the proper planning and conducting of experiments. First, possible confounding variables must be controlled for, usually through the use of comparison. Second, chance should be used in assigning which subjects are to be placed in which groups for which treatment. Third, natural varia-tion in outcomes can be lessened by using more subjects.*

An experiment is performed on objects called *experimental units;* if the units are people, they are called *subjects.* The experimental units or subjects are typically divided into at least two groups. A comparison is made between the response noted in the treatment group or groups and the response noted in the *control group,* a group that receives no treatment (or receives a placebo).

To help minimize the effect of *lurking variables,* it is important to use *ran-domization,* that is, to use chance in deciding which subjects go into which group. It is not sufficient to try to systematically match charac-teristics between the groups, because there are usually variables that one doesn't think of considering until after the results of the experiment start coming in. The best method to use is randomization employing a computer, a hat with names in it, or a random number table.

To help minimize hidden bias, it is often best if subjects do not know which treatment they are receiving. This is called *blinding.* Another precaution is the use of *double-blinding,* in which neither the subjects nor those evaluating their responses know who is receiving which treatment.

KEY EXAMPLE

There is a pressure point on the wrist that some doctors believe can be used to help control the nausea experienced following certain medical procedures. The idea is to place a band containing a small marble firmly on a patient's wrist so that the marble is directly over the pressure point. Describe how an experiment might be run on 50 postoperative patients.

Answer: Assign each patient a number from 01 to 50. From a random number table read off two digits at a time, throwing away repeats and numbers over 50, until 25 numbers have been selected. Put wristbands with marbles over the pressure point on the patients with these selected numbers. Put wristbands with marbles on the remaining patients also, but *not* over the pressure point. Have a researcher check by telephone with all 50 patients at designated intervals to determine the degree of nausea being experienced. Neither the patients nor the researcher on the telephone should know which patients have the marbles over the correct pressure point.

KEY EXAMPLE

A chemical fertilizer company wishes to test whether using their product results in superior vegetables. After dividing a large field into small plots, how might the experiment proceed?

Answer: If the company has one recommended fertilizer application level, half the plots can be randomly selected (assigning the plots numbers and using a random number table) to receive the prescribed dosage of fertilizer. (If the researchers also wish to consider *level*—that is, the amount of fertilizer—randomization should be used for more groupings.) The random selection of plots is to ensure that neither fertilized nor unfertilized plants are inadvertently given land with better rainfall, sunshine, soil type, and so on. To avoid possible bias on the part of employees who will weed and water the plants, they should not know which plots have received the fertilizer. It might be necessary to have containers, one for each plot, of a similar-tasting, similar-smelling substance, half of which contain the fertilizer and the rest containing a chemically inactive material. Finally, if the vegetables are to be judged by quantity and size, the measurements will be less subject to bias. However, if they are to be judged qualitatively, for example, by taste, the judges should not know which vegetables were treated with the fertilizer and which were not.

Two treatments are sometimes compared based on the responses of paired subjects, one of whom receives one treatment while the other receives a second treatment. Often the paired subjects are really single subjects who are given both treatments, one at a time.

KEY EXAMPLE

The famous Pepsi-Coke tests had subjects compare the taste of samples of each drink. How could such a *paired comparison* test be set up?

Answer: It is crucial that such a test be blind, that is, that the subjects not know which cup contains which drink. Furthermore, which drink the subjects taste first should be decided by chance. For example, as each subject arrives, the researcher could read off the next digit from a random number table, with the subject receiving Pepsi or Coke first depending on whether the digit is odd or even.

When differences are observed in a comparison test, the researcher must decide whether these differences are *statistically significant* or whether they can be explained by natural variation. One important consideration is the size of the sample—the larger the sample, the more significant the observation. This is the principle of *replication*.

Just as stratification in sampling design first divides the population into representative groups called *strata, blocking* in experiment design first divides the population into representative groups called *blocks*. This technique helps control certain lurking variables by bringing them directly into the picture and helps make conclusions more specific.

A major goal of experiments is to be able to *generalize* the results to broader populations. Often an experiment must be repeated in a variety of settings. For example, it is hard to generalize from the effect a television commercial has on students at a private midwestern high school to the effect the same commercial has on retired senior citizens in Florida.

Key 39 Theme exercises with answers

OVERVIEW *Sample questions of the type that might appear on homework assignments and tests are presented with answers.*

- What fault do all these sampling designs have in common?

    I. The *Wall Street Journal* plans to make a prediction for a presidential election based on a survey of its readers.

    II. A radio talk show asks people to phone in their views on whether the United States should pay off its huge debt to the United Nations.

    III. A police detective, interested in determining the extent of drug use by teenagers, randomly picks a sample of high school students and interviews each one about any illegal drug use during the past year.

    Answer: All the designs can lead to strong *bias.* The *Wall Street Journal* survey has strong *selection bias;* that is, people who read the *Journal* are not very representative of the general population. The talk show survey results in a *voluntary response sample,* which typically gives too much emphasis to people with strong opinions. The police detective's survey has strong *response bias* in that students may not give truthful responses to a police detective about their illegal drug use.

- A questionnaire is being designed to determine whether most people are or are not in favor of legislation protecting the habitat of the spotted owl. Give two examples of poorly worded questions, one biased toward each response.

    Answer: There are many possible examples, such as "Are you in favor of protecting the habitat of the spotted owl, which is almost extinct and desperately in need of help from an environmentally conscious government?" and "Are you in favor of protecting the habitat of the spotted owl no matter how much unemployment and resulting poverty this causes among hardworking loggers?"

- A study is made to determine whether studying Latin helps students achieve higher scores on the verbal section of the SAT. In comparing records of 200 students, half of whom have taken at least 1 year of Latin, it is noted that the average SAT verbal score is higher for those 100 students who have taken Latin than for those who have not. Based on this study, are the guidance counselors justified in recommending Latin for students who want to do well on the SAT?

Answer: No. Although this study indicates a relation, it does not prove causation, and there could well be a confounding variable responsible for the seeming relationship. For example, it may be that very bright students are the same ones who both choose to take Latin and who do well on the SAT. If students could be randomly assigned to take or not take Latin, and then their SAT scores compared, the results would be more meaningful. Of course, ethical considerations might make it impossible to isolate the confounding variable in this way.

- Explain how you would design an experiment to evaluate whether praying for a hospitalized heart attack patient leads to a speedier recovery.

Answer: For each new heart attack patient entering the hospital, look at the next digit from a random number table. If it is odd, give the name to a group of people who will pray for the patient throughout his or her hospitalization; if it is even, don't give the group the name (randomization). Don't let the patients know what is happening (blinding), and don't let the doctors know what is happening (double-blinding). Compare the lengths of hospitalization of patients who receive prayers with those of control group patients who don't receive prayers (comparison).

- Two studies are run to measure the health benefits of longtime use of daily high doses of vitamin C. Researchers in the first study send a questionnaire to all 50,000 subscribers to a health magazine, asking whether they have taken large doses of vitamin C for at least a 2-year period and what they perceive to be the health benefits, if any. The response rate is 80%. The 10,000 people who did not respond to the first mailing receive follow-up telephone calls, and eventually responses are registered from 98% of the magazine subscribers. Researchers in a second study take a group of 200 volunteers and randomly select 100 to receive high doses of vitamin C while the others receive a similar-looking, similar-tasting placebo. The volunteers are not told whether they are receiving the vitamin, but their doctors know and are asked to note health changes during the 2-year period. Comment on the designs of the two studies, remarking on their good points and on possible sources of error.

Answer: The first study, an observational study, does not suffer from nonresponse bias, as do most mailed questionnaires, because it involved follow-up phone calls and achieved a high response rate. However, this study suffers terribly from selection bias because people who subscribe to a health magazine are not representative of the general population. One would expect most of them to strongly believe that vitamins improve their health. The second study, a controlled experiment, used comparison between a treatment group and a control group, used randomization in selecting who went into each group, and used blinding to control for a placebo effect on the part of the volunteers. However, it did not use double-blinding; that is, the doctors knew whether their patients were receiving the vitamin, and this could have introduced hidden bias when they made judgments about their patients' health.

Theme 6   THE POPULATION PROPORTION

*T*he point of view of statistics is illustrated by the following example. Suppose a jar is filled with black and white balls. If we knew the exact number of balls of each color, we could apply probability theory to predict the likelihood of drawing a ball of a particular color or of drawing a sample of a particular mixture. On the other hand, suppose we do not know the composition of the whole jar, but we have drawn a sample. Statistics tells us with what confidence we can estimate the composition of the whole jar from that of the sample.

In our daily lives we frequently make judgments on the basis of observing the characteristics of samples. A consumer looks at a few sales slips and concludes that a clothing shop is a high- or a low-priced store. A high school senior talks with a group of students and then reaches an opinion about the college these students represent. A traveling salesman rings doorbells for a day and then makes an estimate as to the profitability of a new territory. In all these examples complete observations are impossible or at least impracticable. The same is true in more structured situations. A medical researcher can't test everyone who has a particular form of cancer, a retailer can't test every flashbulb from a shipment (there would be none left to sell!), and a pollster can't survey every potential voter.

Thus we use samples to make inferences about characteristics of the whole population. Many questions arise. With what size and in what manner should a sample be chosen? What conclusions about the population can be drawn from the sample? With what degree of confidence can these conclusions be stated? In answering these questions we keep in mind that we are usually considering only a few members out of a large population and thus can never make any inference about the whole population with 100% certainty. We can, however, draw inferences with specified degrees of certainty.

Key 40 The distribution of sample proportions

OVERVIEW *Whereas the mean is basically a quantitative measurement, the proportion is more of a qualitative approach. The interest is simply in the presence or absence of some attribute. We count the number of "yes" responses and form a proportion. For example, what proportion of drivers wear seat belts? What proportion of SCUD missiles can be intercepted? What proportion of new stereo sets have a certain defect? This separation of the population into "haves" and "have-nots" suggests that we can make use of our earlier work on binomial distributions.*

In this theme we are interested in estimating a population proportion p by considering a single sample proportion $\hat{p}$. This sample proportion is just one out of a whole universe of sample proportions, and to judge its significance we must know how sample proportions vary. Consider the set of proportions from all possible samples of a specified size n. It seems reasonable that these proportions will cluster around the population proportion, and that the larger the chosen sample size, the tighter will be the clustering.

How do we calculate the mean and variance of the set of sample proportions? Suppose the sample size is n, and the actual population proportion is p. From the keys on binomial distributions, we remember that the mean and standard deviation for the number of successes in a given sample are pn and $\sqrt{np(1-p)}$, and for large n the complete distribution begins to look "normal."

We change to proportions by dividing each of these results by n:

$$\mu_{\hat{p}} = \frac{np}{n} = p \text{ and } \sigma_{\hat{p}} = \frac{\sqrt{np(1-p)}}{n} = \sqrt{\frac{p(1-p)}{n}}$$

Thus the principle forming the basis of much of what we do in this theme is as follows:

Start with a population with a given proportion p. Take all samples of size n. Compute the proportion in each of these samples. Then:

- The set of all sample proportions will be approximately *normally* distributed.
- The *mean* of the set of sample proportions will equal *p*, the population proportion.
- The *standard deviation* $\sigma_{\hat{p}}$ of the set of sample proportions will be approximately equal to $\sqrt{p(1-p)/n}$.

KEY EXAMPLE

Suppose that 70% of all dialysis patients will survive at least five years. If 100 new dialysis patients are selected at random, what is the probability that the proportion surviving at least five years will exceed 80%?

Answer: The set of sample proportions is approximately normally distributed with mean .70 and standard deviation

$$\sigma_{\hat{p}} = \sqrt{\frac{(.7)(.3)}{100}} = 0.0458$$

With a *z*-score of (.80 − .70)/0.0458 = 2.18, the probability that our sample proportion exceeds 80% is .5000 − .4854 = .0146.

Key 41 Confidence interval estimate of
the proportion

OVERVIEW *Using a measurement from a sample, we will never be able to say exactly what the population proportion is; rather, we will always say we have a certain confidence that the population proportion lies in a certain interval.*

In finding confidence interval estimates for the population proportion p, since p is unknown, how do we find

$$\sigma_{\hat{p}} = \sqrt{\frac{p(1-p)}{n}}$$

The reasonable procedure is to use the sample proportion $\hat{p}$:

$$\sigma_{\hat{p}} \approx \sqrt{\frac{\hat{p}(1-\hat{p})}{n}}$$

If we have a certain confidence that a sample proportion lies within a specified interval around the population proportion, then we have the same confidence that the population proportion lies within a specified interval about the sample proportion.

Remember that we are really using a normal approximation to the binomial, so $n\hat{p}$ and $n(1-\hat{p})$ should both be at least 10. Furthermore, in making calculations and drawing conclusions from a specific sample, it is important that the sample be a *simple random sample*. Finally, the population should be large, typically checked by the assumption that the sample is less than 10% of the population.

KEY EXAMPLE

If 64% of a sample of 550 people leaving a shopping mall claim to have spent over \$25, determine a 99% confidence interval estimate for the proportion of shopping mall customers who spend over \$25.

Answer: Since $\hat{p} = .64$, the standard deviation of the set of sample proportions is

$$\sigma_{\hat{p}} = \sqrt{\frac{(.64)(.36)}{550}} = 0.0205$$

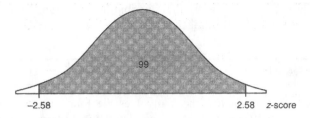

The 99% confidence interval estimate for the population proportion is .64 ± 2.58(0.0205) = .64 ± .053. [On the TI-83, STAT → TESTS → 1-PropZInt. Then x:64∗550, n:550, C-level:.99 and Calculate gives (.58728,.69272).] Thus we are 99% certain that the proportion of shoppers spending over $25 is between .587 and .693.

KEY EXAMPLE

In a random sample of machine parts, 18 out of 225 were found to be damaged in shipment. Establish a 95% confidence interval estimate for the proportion of machine parts that are damaged in shipment.

Answer: The sample proportion is $\hat{p}$ = 18/225 = .08, and the standard deviation of the set of sample proportions is

$$\sigma_{\hat{p}} = \sqrt{\frac{(.08)(.92)}{225}} = 0.0181$$

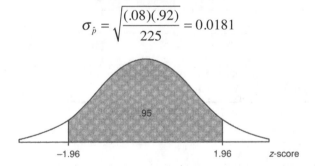

The 95% confidence interval estimate for the population proportion is .08 ± 1.96(0.0181) = .08 ± .035. Thus we are 95% certain that the proportion of machine parts damaged in shipment is between .045 and .115. [On the TI-83, 1-PropZInt gives (.04455,.11545)]

Suppose there are 50,000 parts in the entire shipment. We can translate from proportions to actual numbers: .045(50,000) = 2250 and .115(50,000) = 5750, so we can be 95% confident that there are between 2250 and 5750 damaged parts in the whole shipment.

Some problems relating to the distribution of sample proportions involve one-sided intervals.

KEY EXAMPLE

An assembly-line quality check involves the following procedure. A sample of size 50 is randomly picked, and the machinery is shut down for repairs if the percentage of defective items in the sample is c percent or more. Find the value for c that results in a 90% chance that the machinery will be stopped if, on the average, it is producing 15% of defective items.

Answer: Since $p = .15$,

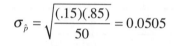

$$\sigma_{\hat{p}} = \sqrt{\frac{(.15)(.85)}{50}} = 0.0505$$

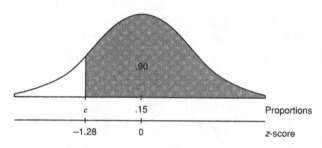

The z-score corresponding to the .90 probability is -1.28, so $c = .15 - 1.28(0.0505) = .085$ or 8.5%.

Key 42 Selecting a sample size

OVERVIEW *One important consideration in setting up surveys is the choice of sample size. To obtain a smaller, more precise interval estimate of the population proportion, we must either decrease the degree of confidence or increase the sample size. Similarly, if we want to increase the degree of confidence, we may either accept a wider interval estimate or increase the sample size. Choosing a larger sample size seems desirable; in the real world, however, this decision involves time and cost considerations.*

In setting up a survey to obtain a confidence interval estimate of the population proportion, what should we use for $\sigma_{\hat{p}}$? It can be shown that $\sqrt{p(1-p)}$ is no larger than .5. Thus $\sqrt{p(1-p)/n}$ is at most $.5/\sqrt{n}$. We make use of this fact in determining sample sizes in problems such as the following.

KEY EXAMPLE

An EPA investigator wants to know the proportion of fish that are inedible because of chemical pollution downstream of an offending factory. If the answer needs to be within ±.03 at the 96% confidence level, how many fish should be in the sample?

Answer:

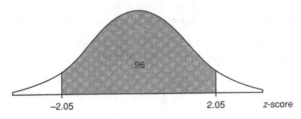

We want $2.05\sigma_{\hat{p}} \le .03$. By the statement given above, $\sigma_{\hat{p}}$ is at most $.5/\sqrt{n}$, so it is sufficient to consider $2.05(.5/\sqrt{n}) \le .03$. Algebraically, we get $\sqrt{n} \ge 2.05(.5)/.03 = 34.17$, and so $n \ge 1167.4$.

Therefore, choosing a sample of 1168 fish will give the inedible proportion to within ±.03 at the 96% level.

Note that the accuracy of the above estimate does *not* depend on what fraction of the whole population we have sampled. What is critical is the absolute size of the sample.

How does the sample size change the accuracy?

Is some minimal value of n necessary for these procedures to be meaningful? Since we are using the normal approximation to the binomial, both np and $n(1 - p)$ should be at least 10 (see Key 33).

KEY EXAMPLE

A study is to be undertaken to determine the proportion of industry executives who believe that workers' pay should be based on individual performance. How many executives should be interviewed if an estimate is desired at the 99% confidence level to within ±0.6? To within ±.03? To within ±.02?

Answer: Algebraically, $2.58(.5 / \sqrt{n}) \le .06$ gives $\sqrt{n} \ge 2.58(.5)/.06 = 21.5$, and so $n \ge 462.25$. Similarly, $2.58(.5/\sqrt{n}) \le .03$ gives $\sqrt{n} \ge 2.58(.5)/.03 = 43$, and so $n \ge 1849$. Finally, $2.58(.5/\sqrt{n}) \le .02$ gives $\sqrt{n} \ge 2.58(.5)/.02 = 64.5$, and so $n \ge 4160.25$. Thus 463, 1849, or 4161 executives should be interviewed depending upon the accuracy desired.

Note that to cut the interval estimate in half (from ±.06 to ±.03), we would have to increase the sample size fourfold (from 462.25 to 1849). To cut the interval estimate to a third (from ±.06 to ±.02), we would have to increase the same size ninefold (from 462.25 to 4160.25).

More generally, if we want to divide the interval estimate by d without effecting the confidence level, we must increase the sample size by a multiple of d^2.

Key 43 Hypothesis test of the proportion

OVERVIEW *Closely related to the problem of estimating a population proportion is the problem of testing a hypothesis about a population proportion. For example, a travel agency might determine an interval estimate for the proportion of sunny days in the Virgin Islands or, alternatively, might test a tourist bureau's claim about the proportion of sunny days. A major stockholder might ascertain an interval estimate for the proportion of successful contract bids or, alternatively, could test a company spokesperson's claim about the proportion of successful bids. A social scientist could find an interval estimate for the proportion of homeless children who attend school or, alternatively, might test a school board member's claim about the proportion of such children who are still able to go to classes. In each of the above, the researcher must decide whether the interest lies in an interval estimate of a population proportion or in a hypothesis test of a claimed proportion.*

The general testing procedure is to choose a specific hypothesis to be tested, called the *null hypothesis*, pick an appropriate random sample, and then use measurements from the sample to determine the likelihood of the null hypothesis. If the sample statistic is far enough away from the claimed population parameter, we say that there is sufficient evidence to reject the null hypothesis. We show that the null hypothesis is unacceptable by showing that it is improbable.

With regard to proportions, the null hypothesis H_0 is stated in the form of an equality statement about the population proportion (for example, $H_0: p = .37$). There is an *alternative hypothesis,* stated in the form of an inequality (for example, $H_a: p < .37$ or $H_a: p > .37$ or $H_a: p \neq .37$). The testing procedure involves picking a sample and comparing the sample proportion $\hat{p}$ with the claimed population proportion p. The strength of the sample statistic $\hat{p}$ can be gauged through its associated P-value, which is the probability of obtaining a sample statistic as extreme as the one obtained if the null hypothesis is assumed to be true. The smaller the P-value, the more significant the difference between the null hypothesis and the sample results.

Alternatively, we can decide on a *critical value c* to gauge the significance of the sample statistic. If the observed $\hat{p}$ is further from the claimed proportion p than is the critical value, we say there is sufficient evidence to reject the null hypothesis. For example, if H_a: $p < .37$ and if $c = .33$, then a sample $\hat{p} = .35$ would not be sufficient evidence to reject H_0.

There are two types of possible errors: the error of mistakenly rejecting a true null hypothesis and the error of mistakenly failing to reject a false null hypothesis. The α-risk, also called the *significance level* of the test, is the probability of committing a *Type I error* and mistakenly rejecting a true null hypothesis. A *Type II error,* a mistaken failure to reject a false null hypothesis, has associated probability β. There is a different value of β for each possible correct value for the population parameter p. For each β, $1 - \beta$ is called the power of the test against the associated correct value.

The assumptions necessary for this hypothesis test for a proportion include the following: a simple random sample, both np and $n(1 - p)$ are at least 10, and the sample size is less than 10% of the population.

Key 44 Critical values, α-risks, and *P*-values

OVERVIEW *We test a null hypothesis by picking a sample and comparing the sample proportion with the claimed population proportion. If the resulting P-value of the test is small enough, we say there is evidence to reject the null hypothesis. Alternatively, we can choose a critical value, or calculate a critical value given an acceptable α-risk, and note if the sample proportion is further from the claimed proportion than is the critical value. We assume we have simple random samples of size less than 10% of the population and that both np and n(1–p) are more than 10.*

KEY EXAMPLE

A local restaurant owner claims that only 15% of visiting tourists stay for more than 2 days. A chamber of commerce volunteer is sure that the real percentage is higher. He plans to survey 100 tourists and intends to speak up if at least 18 of the tourists stay for over 2 days. What is the probability of a Type I error?

Answer:

$$H_0: p = .15, \quad H_a: p > .15$$

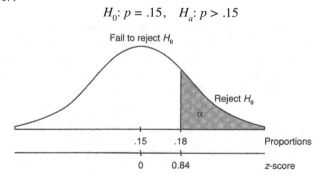

Using the claimed 15%, we calculate the standard deviation of sample proportions to be

$$\sigma_{\hat{p}} = \sqrt{\frac{(.15)(.85)}{100}} = 0.0357$$

The critical proportion is $c = 18/100 = .18$, so the critical z-score is $(.18 - .15)/0.0357 = 0.84$. Thus $\alpha = .5000 - .2995 = .2005$. [On the TI-83, normalcdf(.18, 1, .15, .0357) = .20036.] The test, as set up by the volunteer, has a 20.05% chance of mistakenly rejecting a true null hypothesis.

KEY EXAMPLE

A union spokesperson claims that 75% of the union members support a strike if their basic demands are not met. A company negotiator believes the true percentage is lower and runs a hypothesis test at the 10% significance level. What is the conclusion if 87 of 125 union members say they will strike?

Answer:

$$H_0: p = .75, \quad H_a: p < .75$$

We use the claimed proportion to calculate the standard deviation of the sample proportions.

$$\sigma_{\hat{p}} = \sqrt{\frac{(.75)(.25)}{125}} = 0.03873$$

The observed sample proportion is $\hat{p} = 87/125 = .696$. To measure the strength of the disagreement between the sample proportion and the claimed proportion, we calculate the P-value. The z-score for .696 is $(.696 - .75)/.03873 = -1.39$, with a resulting P-value of .0823. [On the TI-83, normalcdf($-10, -1.39$) = .0823 or normalcdf(0, .696, .75, .03873) = .0816.] Since $.0823 < .10$, there is sufficient evidence to reject H_0 at the 10% significance level. The company negotiator should challenge the union claim at this level. Note, however, that there is not sufficient evidence to reject H_0 at the 5% significance level because $.0823 > .05$. When $.05 < P < .10$, we usually say there is *some* evidence against H_0.

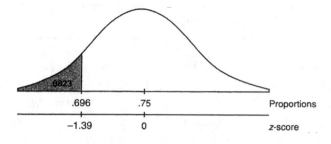

.696 .75 Proportions

−1.39 0 z-score

[On the TI-83, STAT → TESTS → 1PropZTest, then with Po:.75, x:87, n:125, and prop<Po, Calculate gives $P = .0816$.]

KEY EXAMPLE

A cancer research group surveys 500 women over 40 years old to test the hypothesis that 28% of this age group have regularly scheduled mammograms. Should the hypothesis be rejected at the 5% significance level if 151 of the women respond affirmatively?

Answer: Since no suspicion has been voiced that the 28% claim is low or high, we run a two-sided (also called two-tailed) test.

$$H_0: p = .28, \quad H_a: p \neq .28, \quad \alpha = .05$$

$$\sigma_{\hat{p}} = \sqrt{\frac{(.28)(.72)}{500}} = 0.0201$$

The observed $\hat{p} = 151/500 = .302$. The z-score for .302 is $(.302-.28)/.0201 = 1.09$, which corresponds to a probability of .1379 in the tail. Doubling this value (because the test is two-sided), we obtain a P-value of $2(.1379) = .2758$. Because $.2758 > .05$, there is not sufficient evidence to reject H_0; that is, the cancer research group should not dispute the 28% claim. When $P > .10$, we usually say there is little or no evidence to reject H_0. [On the TI-83, 1-PropZTest gives $P = .2732$.]

Key 45 Type II errors and power

OVERVIEW *Why not always choose α to be extremely small to eliminate the possibility of mistakenly rejecting a correct null hypothesis? The difficulty is that this choice would simultaneously increase the chance of never rejecting the null hypothesis even if it were far from true. Thus we see that, for a more nearly complete picture, we must also calculate the probability β of mistakenly failing to reject a false null hypothesis. We shall see that there is a different β-value for each possible correct value for the population proportion p. The operating characteristic curve, or OC curve, a graphical display of β-values, is often given in the real-life analysis of a hypothesis test.*

KEY EXAMPLE

A soft-drink manufacturer received a 9% share of the market this past year. The marketing research department plans a telephone survey of 3000 households. If less than 8% indicate they will buy the company's product, the research department will conclude that the market share has dropped and will order special new promotions. What is the probability of a Type I error?

Answer:

$$H_0: p = .09, \quad H_a: p < .09$$

$$\sigma_{\hat{p}} = \sqrt{\frac{(.09)(.91)}{3000}} = 0.005225$$

The z-score of .08 is $(.08 − .09)/0.005225 = −1.91$, so $\alpha = .5 − .4719 = .0281$.

What is the probability of a Type II error if the true market share is .085? In other words, if the market share really has dropped (to 8.5%), what is the probability that the research department will mistakenly fail to reject the 9% null hypothesis?

Answer: The z-score of .08 now is $(.08 − .085)/0.005225 = −0.96$. Thus $\beta = .5000 + .3315 = .8315$.

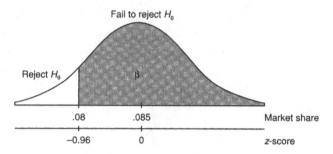

What if the true market share is .075?

Answer: The z-score of .08 now is $(.08 − .075)/0.005225 = 0.96$. Thus $\beta = .5000 − .3315 = .1685$.

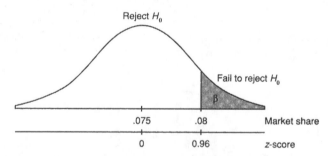

Note that the further the true value is away (in the suspected direction) from the claimed null hypothesis, the smaller the probability is of failing to reject the false claim.

A building inspector believes that the percentage of new construction with serious code violations may be even greater than the previously claimed 7%. A hypothesis test is planned on 200 new homes at the 1% significance level. What is the β-value if the true percentage of new constructions with serious violations is 9%? Is 11%? Is 13%?

Answer:

$$H_0: p = .07, H_a: p > .07, \quad a = .01$$

$$\sigma_{\hat{p}} = \sqrt{\frac{(.07)(.93)}{200}} = 0.018$$

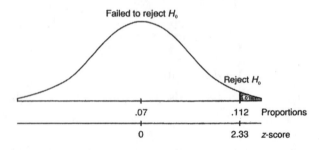

The 1% significance level gives a critical z-score of 2.33 and a critical proportion of .07 + 2.33(0.018) = .112.

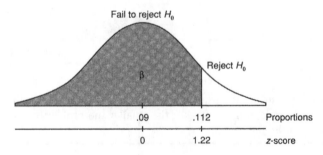

If the true percentage is 9%, the z-score of .112 is (.112 − .09)/0.018 = 1.22. Then β = .5000 + .3888 = .8888.

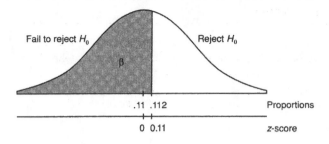

If the true percentage is 11%, the z-score of .112 is $(.112 - .11)/0.018 = 0.11$. Then $\beta = .5000 + .0438 = .5438$.

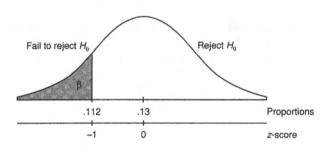

If the true percentage is 13%, the z-score of .112 is $(.112 - .13)/0.018 = -1$. Then $\beta = .5000 - .3413 = .1587$.

A table of these values (and a few more) for β is as follows:

True p	.08	.09	.10	.11	.12	.13	.14	.15
β	.9625	.8888	.7486	.5438	.3300	.1587	.0594	.0174

The resulting graph is called the *operating characteristic (OC) curve*.

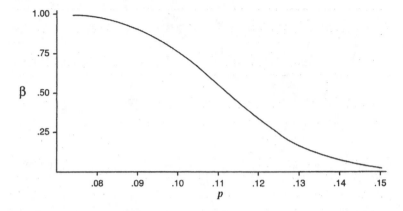

If β is the probability of failing to reject a false null hypothesis, then 1 − β is the probability of rejecting the false null hypothesis. The *power* of a hypothesis test is the probability that a Type II error is not committed, and the graph of 1 − β is called the *power curve*.

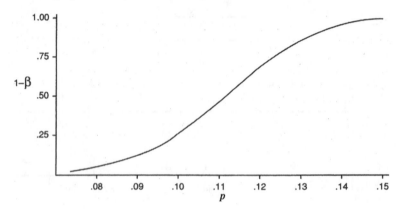

Both the operating characteristic curve and the related power curve help one to gauge the effectiveness of a hypothesis test.

Key 46 Confidence interval estimate for
the difference of two proportions

OVERVIEW *Numerous important and interesting applications of statistics involve the comparison of two population proportions. For example, is the proportion of satisfied purchasers of American automobiles greater than that for buyers of Japanese cars? How does the percentage of surgeons recommending a new cancer treatment compare with the corresponding percentage of oncologists? What can be said about the difference between the proportion of single parents who are on welfare and the proportion of two-parent families on welfare?*

Often it is clear that some difference exists between two population proportions, and we would like to give a numerical measure of the difference. Using samples, we cannot give this difference *exactly;* however, we can say, with a specified degree of confidence, that the difference lies in a certain interval. We follow the same procedure as set forth in Key 41, this time using

$$p_1 - p_2, \quad \sqrt{\frac{p_1(1-p_1)}{n_1} + \frac{p_2(1-p_2)}{n_2}}, \quad \text{and} \quad \hat{p}_1 - \hat{p}_2$$

in place of

$$p, \quad \sqrt{\frac{p(1-p)}{n}}, \quad \text{and} \quad \hat{p}$$

As in the preceding keys on proportions, we are using the normal to estimate the binomial and so will assume that

$$n_1 p_1, \quad n_1(1-p_1), \quad n_2 p_2, \quad \text{and} \quad n_2(1-p_2),$$

are all at least 10. We also assume both that the samples are *simple random samples* and that they are taken *independently* of each other, and that the original populations are large compared with the sample sizes.

KEY EXAMPLE

Suppose that 84% of a sample of 125 nurses working 7 A.M. to 3 P.M. shifts in city hospitals express positive job satisfaction, while only 72% of a sample of 150 nurses on 11 P.M. to 7 A.M. shifts express similar fulfillment. Establish a 90% confidence interval estimate for the difference.

Answer:

$$n_1 = 125, \quad n_2 = 150$$
$$\hat{p}_1 = .84, \quad \hat{p}_2 = .72$$
$$\sigma_d = \sqrt{\frac{(.84)(.16)}{125} + \frac{(.72)(.28)}{150}} = 0.0492$$

The observed difference is $.84 - .72 = .12$, and the critical z-scores are ± 1.645. The confidence interval estimate is $.12 \pm 1.645(0.0492) = .12 \pm .081$. We can be 90% certain that the proportion of satisfied nurses on 7 to 3 shifts is between .039 and .201 higher than for those on 11 to 7 shifts. [On the TI-83, 2-PropZInt gives (.0391,.2009).]

KEY EXAMPLE

A grocery store manager notes that, in a sample of 85 people going through the "under 7 items" checkout line, only 10 paid with checks; whereas, in a sample of 92 customers passing through the regular line, 37 paid with checks. Find a 95% confidence interval estimate for the difference between the proportions of customers going through the two different lines who use checks.

Answer:

$$n_1 = 85, \quad n_2 = 92$$

$$\hat{p}_1 = \frac{10}{85} = .118, \quad \hat{p}_2 = \frac{37}{92} = .402$$

$$\sigma_d = \sqrt{\frac{(.118)(.882)}{85} + \frac{(.402)(.598)}{92}} = 0.0619$$

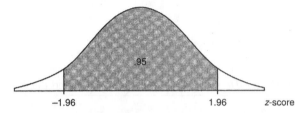

The observed difference is $.118 - .402 = -.284$, and the critical z-scores are ± 1.96. Thus, (see also Key 50) the confidence interval estimate is $-.284 \pm 1.96(0.0619) = -.284 \pm .121$. The manager can be 90% sure that the proportion of customers passing through the "under 7 items" line who use checks is between .163 and .405 lower than the proportion going through the regular line who use checks. [On the TI-83, 2-PropZInt gives $(-.4059, -.1632)$.]

Setting up experiments or surveys involves many considerations, one of which is *sample size*. Generally, if we want smaller, more precise interval estimates, we either decrease the degree of confidence or increase the sample size. Similarly, if we want to increase the degree of confidence, we may either accept a wider interval or again increase the sample size.

In Key 42 we noted that $\sqrt{p(1-p)}$ is at most .5. Thus

$$\sqrt{p(1-p)\left(\frac{1}{n_1} + \frac{1}{n_2}\right)} \leq (.5)\sqrt{\frac{1}{n_1} + \frac{1}{n_2}}$$

Now, if we simplify by insisting that $n_1 = n_2 = n$, the above expression reduces as follows:

$$(.5)\sqrt{\frac{1}{n} + \frac{1}{n}} = (.5)\sqrt{\frac{2}{n}} = \frac{.5\sqrt{2}}{\sqrt{n}}$$

KEY EXAMPLE

A pollster wants to determine the difference between the proportions of high-income voters and of low-income voters who support a decrease in capital gains taxes. If the answer needs to be known to within ±.02 at the 95% confidence level, what size samples should be taken?

Answer: Assuming we will pick the same size samples for the two sample proportions, we have

$$\sigma_d \leq \frac{.5\sqrt{2}}{\sqrt{n}} \quad \text{and} \quad 1.96\sigma_d \leq .02$$

Thus, $1.96(.5)\sqrt{2}/\sqrt{n} \leq .02$, and algebraically we find that

$$\sqrt{n} \geq \frac{1.96(.5)\sqrt{2}}{.02} = 69.3$$

Therefore, $n \geq 69.3^2 = 4802.5$, and the pollster should use 4803 people for each sample.

Key 47 Hypothesis test for the difference
of two proportions

OVERVIEW *To compare proportions from two different populations, we consider a sample from each population and note the difference between the sample proportions. The strength of this difference can be measured by calculating the P-value. As with confidence intervals for the differences of two proportions,* $n_1\hat{p}_1$, $n_1(1 - \hat{p}_1)$, $n_2\hat{p}_2$, *and* $n_2(1 - \hat{p}_2)$ *should all be at least 10, the samples should be **simple random samples** and be taken **independently** of each other, and the original populations should be large compared with the sample sizes.*

We consider problems for which the null hypothesis states that the population proportions are equal, or, equivalently, that their difference is 0:

$$H_0: p_1 - p_2 = 0$$

The alternative hypothesis is then

$$H_a: p_1 - p_2 < 0, \quad H_a: p_1 - p_2 > 0, \quad \text{or} \quad H_a: p_1 - p_2 \neq 0$$

The first two possibilities lead to one-sided (one-tailed) tests, while the third possibility leads to two-sided (two-tailed) tests.

Since the null hypothesis is that $p_1 = p_2$, we call this common value p, and use it to calculate σ_d:

$$\sigma_d = \sqrt{\frac{p(1-p)}{n_1} + \frac{p(1-p)}{n_2}} = \sqrt{p(1-p)\left(\frac{1}{n_1} + \frac{1}{n_2}\right)}$$

In practice, if

$$\hat{p}_1 = \frac{x_1}{n_1} \quad \text{and} \quad \hat{p}_2 = \frac{x_2}{n_2}$$

we use

$$\hat{p} = \frac{x_1 + x_2}{n_1 + n_2}$$

as an estimate for p in calculating σ_d.

KEY EXAMPLE

Suppose, early in an election campaign, a telephone poll of 800 registered voters shows 460 in favor of a particular candidate. Just before election day, a second poll shows 520 out of 1000 registered voters expressing this preference. At the 10% significance level is there sufficient evidence that the candidate's popularity has decreased?

Answer: $H_0: p_1 - p_2 = 0, \quad H_a: p_1 - p_2 > 0, \quad \alpha = .10$

$$\hat{p}_1 = \frac{460}{800} = .575, \quad \hat{p}_2 = \frac{520}{1000} = .520, \quad \hat{p} = \frac{460 + 520}{800 + 1000} = .544$$

$$\sigma_d = \sqrt{(.544)(.456)\left(\frac{1}{800} + \frac{1}{1000}\right)} = 0.0236$$

The observed difference is $.575 - .520 = .055$. The z-score for .055 is $(.055-0)/.0236 = 2.33$, and so the P-value is .0099. [On the TI-83, normalcdf(2.33,100) = .00990 or normalcdf(.055,1,0,,0236) = .00989.] Because $.0099 < .10$, we conclude that at the 10% significance level the candidate's popularity *has* dropped. We note that $.0099 < .01$, so that the observed difference is statistically significant even at the 1% level.

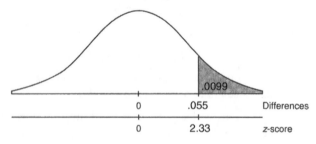

[On the TI-83, 2-PropZTest gives $P = .00995$.]

KEY EXAMPLE

An automobile manufacturer tries two distinct assembly procedures. In a sample of 350 cars coming off the line using the first procedure, there are 28 with major defects, while a sample of 500 autos from the second line shows 32 with defects. Is the difference significant at the 10% significance level?

Answer: Since there is no mention that one of the procedures is believed to be better or worse than the other, this is a two-sided test.

$$H_0: p_1 - p_2 = 0, \quad H_a: p_1 - p_2 > 0, \quad \alpha = .10$$

$$\hat{p}_1 = \frac{28}{350} = .080, \quad \hat{p}_2 = \frac{32}{500} = .064, \quad \hat{p} = \frac{28 + 32}{350 + 500} = .0706$$

$$\sigma_d = \sqrt{(.0706)(.9294)\left(\frac{1}{350} + \frac{1}{500}\right)} = 0.0179$$

The observed difference is $.080 - .064 = .016$. The z-score for $.016$ is $(.016 - 0)/.0179 = 0.89$, which corresponds to a tail probability of $.1867$. [On the TI-83, normalcdf(.89,1000) = .1867 or normalcdf(.016,1,0,.0179) = .1857.] Doubling this value because the test is two-sided results in a P-value of $2(.1867) = .3734$. [On the TI-83, 2-PropZTest gives $P = .3701$.] Because $.3734 > .10$, we conclude that the observed difference is *not* significant at the 10% level. When P is so large, we can safely say that there is no evidence against H_0.

Key 48 Theme exercises with answers

OVERVIEW *Sample questions of the type that might appear on homework assignments and tests are presented with answers.*

- In one study, 18% of 100 people with migraine headaches experienced substantial symptomatic relief after taking a placebo. Establish a 90% confidence interval estimate of the percentage of migraine sufferers who can be helped by placebos.

  Answer: Since $\hat{p} = .18$, the standard deviation of the set of sample proportions is

  $$\sigma_{\hat{p}} = \sqrt{\frac{(.18)(.82)}{100}} = 0.03842$$

  The 90% confidence interval estimate for the population proportion is $.18 \pm 1.645(0.03842) = .18 \pm .063$. Thus we are 90% certain that the proportion of migraine sufferers who can be helped by placebos is between .117 and .243. [On the TI-83, 1-PropZInt gives (.11681,.24319).]

- In a random sample of 75 clothing purchases returned for refunds, the buyers claimed that 47 were brought back because of improper fit. Construct a 94% confidence interval estimate for the proportion of clothing returns that are blamed on fit.

  Answer: The sample proportion is $\hat{p} = 47/75 = .627$, and the standard deviation of the set of sample proportions is

  $$\sigma_{\hat{p}} = \sqrt{\frac{(.627)(.373)}{75}} = 0.0558$$

  The 94% confidence interval estimate for the population proportion is $.627 \pm 1.88(.0558) = .627 \pm .105$. Thus we are 94% certain that between .522 and .732 of the clothing returns are blamed on poor fit. [On the TI-83, 1-PropZInt gives (.52162,.73171).]

- A Department of Labor survey of 6230 unemployed adults classified people by marital status, sex, and race. The raw numbers are as follows:

	White, 16 Years and Over			Nonwhite, 16 Years and Over		
	Married	Widow/Div.	Single	Married	Window/Div.	Single
Men	1090	337	1168	266	135	503
Women	952	423	632	189	186	349

Find a 90% confidence interval estimate for the proportion of unemployed men who are married.

Answer: Totaling the first row across, we find that there were 3499 men in the survey, and we note that $1090 + 266 = 1356$ of these were married. Therefore,

$$\hat{p} = \frac{1356}{3499} = .3875 \quad \text{and} \quad \sigma_{\hat{p}} = \sqrt{\frac{(.3875)(.6125)}{3499}} = 0.008236$$

Thus the 90% confidence interval estimate is $.3875 \pm 1.645 (0.008236) = .3875 \pm .0135$. [On the TI-83, 1-PropZInt gives $(.37399,.40109)$.]

Find a 98% confidence interval estimate of the proportion of unemployed singles who are women.

Answer: There were $1168 + 632 + 503 + 349 = 2652$ singles in the survey, and of these $632 + 349 = 981$ were women. Therefore,

$$\hat{p} = \frac{981}{2652} = 0.3699 \quad \text{and} \quad \sigma_{\hat{p}} = \sqrt{\frac{(.3699)(.6301)}{2652}} = 0.009375$$

The 98% confidence interval estimate is $.3699 \pm 2.33(0.009375) = .3699 \pm .0218$. [On the TI-83, 1-PropZInt gives $(.3481,.39172)$.]

- A telephone survey of 1000 adults was taken shortly after the United States began bombing Iraq in 1990. If 832 voiced their support for this action, with what confidence can it be asserted that $83.2\% \pm 3\%$ of the adult population supported the decision to go to war?

$$\hat{p} = \frac{832}{1000} = .832 \quad \text{and} \quad \sigma_{\hat{p}} = \sqrt{\frac{(.832)(.168)}{1000}} = 0.0118$$

The relevant z-scores are $\pm.03/0.0118 = \pm2.54$. Table A gives a probability of .4945, and so our answer is $2(.4945) = .9890$. In other words, $83.2\% \pm 3\%$ is a 98.90% confidence interval estimate for adult support of the war decision.

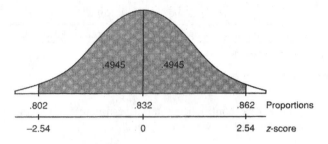

.802	.832	.862	Proportions
−2.54	0	2.54	z-score

If the adult U.S. population is 191,000,000, estimate the actual numerical support.

Answer: Since .802(191,000,000) ≈ 153,000,000, while .862(191,000,000) ≈ 165,000,000, we can be 98.90% sure that between 153 and 165 million adults supported the initial bombing decision.

• The American Medical Association wishes to determine the percentage of obstetricians who are considering leaving the profession because of the rapidly increasing number of lawsuits against obstetricians. How large a sample should be taken to find the answer to within ±2% at the 95% confidence level?

Answer: We want $1.96\sigma_{\hat{p}} \leq .02$. Since $\sigma_{\hat{p}}$ is at most $.5/\sqrt{n}$ it is sufficient to consider $1.96(.5/\sqrt{n}) \leq .02$. Algebraically, we get $\sqrt{n} \geq 1.96(.5)/.02 = 49$, and so $n \geq 2401$. Thus interviewing 2401 obstetricians will give the required proportion to within ±.02 at the 95% level.

• It is believed that 60% of all whiskey drinkers can tell the difference between Rot Gut and Northern Comfort. A hypothesis test is run on 50 pub customers, and 35 are able to distinguish between the whiskeys. Is this sufficient evidence at the 5% significance level to call into question the 60% claim?

Answer:

$$H_0{:}p = .60, \quad H_a{:}p \neq .60, \quad \alpha = .05$$

We use the claimed 60% to calculate the standard deviation of sample proportions to be

$$\sigma_{\hat{p}} = \sqrt{\frac{(.60)(.40)}{50}} = 0.0693$$

The observed $\hat{p}$ = 35/50 = .70. The z-score for .70 is $(.70 − .60)/.0693 = 1.44$, which corresponds to a probability of .0749 in the tail. Doubling this value (because the test is two-sided), we obtain a P-value of $2(.0749) = .1498$. Because $.1498 > .05$, there is not sufficient evidence to reject H_0; that is, the 60% claim should not be disputed. [On the TI-83, 1-PropZTest gives $P = .1489$.]

- A television producer knows that, unless a new show is watched by at least 25% of the possible audience, it will be canceled. Suppose a quick survey finds only 21 out of 100 viewers tuned in to the show. Is this sufficient evidence at the 10% significance level that the show will be canceled? Determine the P-value.

Answer:

$$H_0: p = .25, \quad H_a: p < .25, \quad a = .10$$

$$\sigma_{\hat{p}} = \sqrt{\frac{(.25)(.75)}{100}} = 0.0433$$

The observed $\hat{p} = 21/100 = .21$. The z-score for $.21$ is $(.21 - .25)/.0433 = -0.924$, with a resulting P-value of $.5 - .3212 = .1788$. Because $.1788 > .10$, there is not sufficient evidence to cancel the new show. [On the TI-83, 1-PropZTest gives $P = .1778$.]

- An airline claims that 92% of its flights leave on schedule, but an FAA investigator believes the true figure is lower. He decides that 125 flights will be checked at the 5% significance level. What is the probability of a Type II error if the true percentage is 90%?

Answer:

$$H_0: p = .92, \quad H_a: p < .92, \quad a = .05,$$

$$\sigma_{\hat{p}} = \sqrt{\frac{(.92)(.08)}{125}} = 0.0243$$

With $\alpha = .05$ the critical z-score is 1.645, and the critical proportion is $.92 - 1.645(0.0243) = .880$. If the true proportion of on-schedule flights is $.90$, then the z-score of $.880$ is $(.880 - .90)/0.0243 = 0.82$, and $\beta = .5 + .2939 = .7939$.

- A survey of 5000 medical students compared the career goals of men and women.

Career Goal

	Surgery	Gynecology	Pediatrics	Psychiatry	Other
Men	312	520	472	610	1026
Women	128	350	391	400	791

Establish a 95% confidence interval estimate for the difference between the proportions of men and women intending to become surgeons.

Answer:

$$\hat{p}_1 = \frac{312}{312 + 520 + 472 + 610 + 1026} = \frac{312}{2940} = .106$$

$$\hat{p}_2 = \frac{128}{128 + 350 + 391 + 400 + 791} = \frac{128}{2060} = .062$$

$$\sigma_d = \sqrt{\frac{(.106)(.894)}{2940} + \frac{(.062)(.938)}{2060}} = 0.00778$$

The observed difference is .106 − .062 = .044, so the confidence interval estimate is .044 ± 1.96(0.00778) = .044 ± .015. Therefore we can be 95% certain that the proportion of men heading toward surgery is between .029 and .059 higher than the proportion of women. [On the TI-83, 2-PropZInt gives (.02873, .05924).]

Is the observed difference between the proportions of men and women hoping to become gynecologists significant at the 2% significance level?

Answer:

$$H_0\colon p_1 - p_2 = 0, \quad H_a\colon p_1 - p_2 \neq 0, \quad a = .02$$

$$\hat{p}_1 = \frac{520}{2940} = .177, \quad \hat{p}_2 = \frac{350}{2060} = .170, \quad \hat{p} = \frac{520 + 350}{5000} = .174$$

$$\sigma_d = \sqrt{(.174)(.826)\left(\frac{1}{2940} + \frac{1}{2060}\right)} = 0.0109$$

The observed difference is .177 − .170 = .007. The z-score for .007 is (.007 − 0)/.0109 = 0.64, which corresponds to a tail probability of .2611. [On the TI-83, normalcdf(.64,1000 = .2611 or normalcdf(.007,1,0,.0109) = .2604.] Doubling this value because the test is two-sided results in a *P*-value of 2(.2611) = .5222. [On the TI-83, 2-PropZTest gives *P* = .5224.] Because .5222 > .02, we conclude that the observed difference between the proportions of men and women looking toward gynecology is *not* significant at the 2% level.

Is there sufficient evidence at the 0.5% significance level that a higher proportion of women want to become pediatricians than do men?

Answer:

$$H_0: p_1 - p_2 = 0, \quad H_a: p_1 - p_2 > 0, \quad a = .005,$$

$$\hat{p}_1 = \frac{391}{2060} = .190, \quad \hat{p}_2 = \frac{472}{2940} = .161, \quad \hat{p} = \frac{391 + 472}{5000} = .173$$

$$\sigma_d = \sqrt{(.173)(.827)\left(\frac{1}{2060} + \frac{1}{2940}\right)} = 0.0109$$

The observed difference is $.190 - .161 = .029$. The z-score for .029 is $(.029 - 0)/.0109 = 2.66$, which gives a P-value of .0039. [On the TI-83, normalcdf(2.66,1000) = .0039 or normalcdf(.029,1,0,.0109) = .0039.] Because $.0039 < .005$, there is sufficient evidence that a greater proportion of women than men plan to become pediatricians. [On the TI-83, 2-PropZTest gives $P = .0035$.]

Theme 7 THE POPULATION MEAN

*T*heme 6 dealt with estimates and tests for a population *proportion*. Equally important in numerous applications are techniques and procedures involving a population *mean*. Whereas the proportion essentially represents a qualitative approach, the mean is basically a quantitative measurement. For example, what is the mean salary of computer programmers (rather than what proportion of the salaries are above $50,000)? What is the mean height of basketball players (rather than what is the proportion of players over 6 feet 5 inches tall)? What is the mean number of new policies sold each week (rather than what is the proportion of weeks for which at least ten new policies are sold)?

Just as with proportions, it is usually impossible or impractical to gather complete information about means. Thus we must use techniques to draw inferences about a population mean when only a sample mean is available. As we should now expect, results will involve confidence intervals and degrees of certainty; that is, we cannot make any inference about the population mean with 100% certainty. Confidence interval estimates and hypothesis tests for both single-population means and differences of two means will be examined.

INDIVIDUAL KEYS IN THIS THEME

Key 49 The distribution of sample means

OVERVIEW *Heights, weights, and the like tend to result in normal distributions, but normally distributed natural phenomena occur much less frequently than one might guess. However, the normal curve has an importance in statistics that is independent of whether or not it appears in nature. What is significant is that the results of many types of sampling experiments can be analyzed using the normal curve. For example, there is no reason to suppose that the amounts of money that different people spend in grocery stores are normally distributed. However, if everyday we survey 30 people leaving a store and determine the average grocery bill, then these daily averages will have a nearly normal distribution. This statistical key will enable us to use a sample mean to estimate a population mean.*

We are interested in estimating the mean of a population. For our estimate we could simply randomly pick a single element of the population, but we then would have little confidence in our answer. Suppose instead we pick 100 elements and calculate their average. It is intuitively clear that the resulting sample mean has a greater chance of being closer to the mean of the whole population than does the value for any individual member of the population.

When we pick a sample and measure its mean, we are finding exactly one sample mean out of a whole universe of sample means. To judge the significance of a single sample mean, we must know how sample means vary. Consider the set of means from all possible samples of a specified size. It is both apparent and reasonable that the sample means will be clustered about the mean of the whole population; furthermore, these sample means will have a tighter clustering than do the elements of the original population. In fact, we might guess that the larger the chosen sample size, the tighter the clustering will be.

The following principle forms the basis of much of what we do in following keys. It is a simplified statement of the *central limit theorem* of statistics.

Start with a population with a given mean μ and standard deviation σ. Pick *n* sufficiently large (at least 30), and take all samples of size *n*. Compute the mean of each of these samples. Then:

- The set of all sample means will be approximately *normally* distributed.
- The *mean* of the set of sample means will equal μ, the mean of the population.
- The *standard deviation,* $\sigma_{\bar{x}}$, of the set of sample means will be approximately equal to $\sigma/\sqrt{n}$, that is, equal to the standard deviation of the whole population divided by the square root of the sample size.

KEY EXAMPLE

Suppose that the average outstanding credit card balance among young couples is $650 with a standard deviation of $420. If 100 couples are selected at random, what is the probability that the mean outstanding credit card balance exceeds $700?

Answer: The sample size is over 30, so by the central limit theorem, the set of sample means is approximately normally distributed with mean 650 and standard deviation $420/\sqrt{100}$ = 42. With a *z*-score of (700 − 650)/42 = 1.19, the probability of our sample mean exceeds 700 is .5000 − .3830 = .1170. [On the TI-83, normalcdf(700, 10000, 650, 42) = .1169.]

KEY EXAMPLE

The strength of paper coming from a manufacturing plant is known to be 25 pounds per square inch with a standard deviation of 2.3. In a simple random sample of 40 pieces of paper, what is the probability that the mean strength is between 24.5 and 25.5 pounds per square inch?

Answer: $\mu_{\bar{x}}$ = 25 and $\sigma_{\bar{x}} = 2.3/\sqrt{40}$ = 0.364. The *z*-scores of 24.5 and 25.5 are (24.5 − 25)/0.364 = −1.37 and (25.5 − 25)/0.364 = 1.37, respectively. The probability that the mean strength in the sample is between 24.5 and 25.5 pounds per square inch is .4147 + .4147 = .8294. [On the TI=83, normalcdf(24.5,25.5,25,.364) = .8304.]

Key 50 Student *t*-distributions

OVERVIEW *When the population standard deviation σ is unknown, we use the sample standard deviation s as an estimate for σ, and we use $s/\sqrt{n}$ as an estimate for $\sigma_{\bar{x}} = \sigma/\sqrt{n}$. In this situation, we work with the variable $t = \dfrac{\bar{x} - \mu}{\sigma_{\bar{x}}}$ with a resulting t-distribution that is bell-shaped and symmetric, but lower at the mean, higher at the tails, and so more spread out than the normal distribution.*

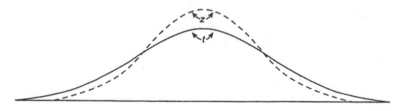

Like the binomial distribution, the *t*-distribution is different for different values of *n*. In tables these distinct *t*-distributions are associated with the value of df (degrees of freedom), which in our present consideration is equal to the sample size minus 1. The smaller the df value, the larger is the dispersion in the distribution. The larger the df value, that is, the larger the sample size, the closer the distribution comes to the normal distribution.

Since there is a separate *t*-distribution for each degree of freedom, fairly complete tables would involve many pages; therefore, we give areas and *t*-values for only the more commonly used percentages or probabilities. The last row of Table B in the Appendix is the normal distribution because the normal distribution is a special case of the *t*-distribution taken when *n* is infinite. For practical purposes, the two distributions are very close for any $n \geq 30$.

Note that, whereas Table A gives areas under the normal curve from the mean to positive *z* values, Table B gives areas to the right of given positive *t* values. For example, suppose the sample size is 20 and so df = 20 – 1 = 19. Then a probability of .05 in the *tail* will correspond with a *t* value of 1.729, while .01 in the tail corresponds to *t* = 2.539.

Thus the t-distribution is the proper choice whenever the population standard deviation σ is unknown. In the real world σ is almost always unknown, so we should almost always use the t-distribution. In the past this was difficult because extensive tables were necessary for the various t-distributions. However, with calculators such as the TI-83, this problem has diminished. It is no longer necessary to assume that the t-distribution is close enough to the z-distribution whenever n is greater than the arbitrary number 30.

Use of the t-distribution with a simple random sample of size n is refined even further by some statisticians: If n is large (≥ 40), it is unnecessary to make any assumptions about the parent population; if n is medium ($15 - 40$), then either the sample should show no extreme values and little skewness or the parent population must be assumed normal; and if n is small (≤ 15), then either the sample should show no outliers and no skewness or the parent population must be assumed normal.

Key 51 Confidence interval estimate of the mean

OVERVIEW *Using a measurement from a sample, we will never be able to say exactly what the population mean is; rather, we will always say we have a certain confidence that the population mean lies in a certain interval.*

KEY EXAMPLE

A bottling machine is operating with a standard deviation of 0.12 ounce. Suppose that in a sample of 36 containers the machine dumped an average of 16.1 ounces. Estimate the mean number of ounces in all bottles that this machine fills. More specifically, give an interval in which we are 95% certain that the mean lies.

Answer: For samples of size 36, the sample means are approximately normally distributed with a standard deviation of

$$\sigma_{\bar{x}} = \sigma / \sqrt{n} = 0.12 / \sqrt{36} = 0.02$$

From Key 29 we have that 95% of the sample means should be within 1.96 standard deviations of the population mean. Equivalently, we are 95% certain that the population mean is within 1.96 standard deviations of any sample mean.* In our case, $16.1 \pm 1.96(0.02) = 16.1 \pm 0.0392$, and we are 95% sure that the mean number of ounces in all bottles is between 16.0608 and 16.1392. This is called a *95% confidence interval estimate*. [On the TI-83, STAT $\rightarrow$ TESTS $\rightarrow$ ZInterval, with Inpt:Stats, σ:12, $\bar{x}$:16.1, n:36, and C-Level:.95, Calculate gives (16.061,16.139).]

How about a 99% confidence interval estimate?

Answer: Here, $16.1 \pm 2.58(0.02) = 16.1 \pm 0.0516$, and we are 99% sure that the mean number of ounces in all bottles is between 16.0484 and 16.1516. [On the TI-83, ZInterval gives (16.048,16.152).]

---

*Note that we cannot say there is a .95 *probability* that the population mean is within 1.96 standard deviations of a given sample mean. For a given sample mean, the population mean either is or isn't within the specified interval, so the probability is either 1 or 0.

Note that, when we wanted a higher certainty (99% instead of 95%), we had to settle for a larger, less specific interval (±0.0516 instead of ±0.0392).

Remember, when σ is unknown (which is almost always the case), we use the *t*-distribution instead of the *z*-distribution, and if the sample is small, we assume the parent population is approximately normal.

KEY EXAMPLE

When ten cars of a new model were tested for gas mileage, the results showed a mean of 27.2 mpg with a standard deviation of 1.8 mpg. A 95% confidence interval estimate for the miles per gallon achieved by this model is obtained as follows:

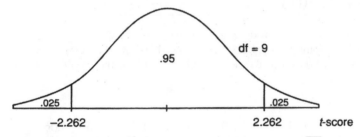

The standard deviation of the sample means $\sigma_{\bar{x}} = 1.8/\sqrt{10} = 0.569$. With $10 - 1 = 9$ degrees of freedom, and 2.5% in each tail, the appropriate *t*-scores are ±2.262. Thus we can be 95% certain that the gas mileage of the new model is in the range $27.2 \pm 2.262(0.569) = 27.2 \pm 1.3$, or between 25.9 and 28.5 mpg. [On the TI-83, STAT → TESTS → TInterval, with Stats, $\bar{x}$:27.2, Sx:1.8, n:10, and C-Level:.95, Calculate gives (25.912, 28.488).]

KEY EXAMPLE

Twenty-five "18 ounce" jars of peanut butter are weighed, yielding the totals $\Sigma x = 448.5$ and $\Sigma(x - \bar{x})^2 = 0.41$. What is the 99% confidence interval estimate for the mean weight?

Answer: We first calculate the sample mean and standard deviation:

$$\bar{x} = \frac{\Sigma x}{n} = \frac{448.5}{25} = 17.94, \quad s = \sqrt{\frac{\Sigma(x - \bar{x}^2}{n-1}} = \sqrt{\frac{0.41}{24}} = 0.13$$

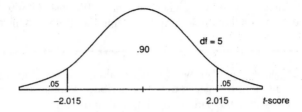

The standard deviation $\sigma_{\bar{x}}$ of the sample means is estimated to be $s/\sqrt{n} = 0.13/\sqrt{25} = 0.026$. With $25 - 1 = 24$ degrees of freedom, and 0.5% in each tail, the appropriate t-scores are ±2.797. Thus a 99% confidence interval estimate is given by $17.94 \pm 2.797(0.026) = 17.94 \pm 0.073$ ounces. [On the TI-83, TInterval gives (17.867, 18.013).]

KEY EXAMPLE

A new process for producing synthetic gems yielded, in its first run, six stones weighing 0.43, 0.52, 0.46, 0.49, 0.60, and 0.56 carat, respectively. Find a 90% confidence interval estimate for the mean carat weight from this process.

Answer:

$$\bar{x} = \frac{\sum x}{n} = \frac{0.43 + 0.52 + 0.46 + 0.49 + 0.60 + 0.56}{6} = \frac{3.06}{6} = 0.51$$

$$s = \sqrt{\frac{\sum(x - \bar{x})^2}{n-1}}$$

$$= \sqrt{\frac{(0.08)^2 + (0.01)^2 + (0.05)^2 + (0.02)^2 + (0.09)^2 + (0.05)^2}{5}}$$

$$= 0.0632$$

$$\sigma_{\bar{x}} = \frac{s}{\sqrt{n}} = \frac{0.0632}{\sqrt{6}} = 0.0258$$

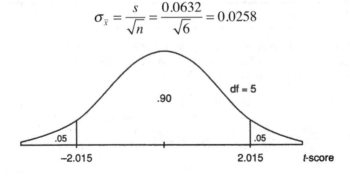

With df = 6 − 1 = 5, and 5% in each tail, the t-scores are ±2.015. Thus we can be 90% sure that the new process will yield stones weighing 0.51 ± 2.015(0.0258) = 0.51 ± 0.052, or between 0.458 and 0.562 carat. [On the TI-83, put the data in a list and then use Data under TInterval to obtain (.45797, .56203).]

Key 52 Selecting a sample size

OVERVIEW *Statistical principles are useful not only in analyzing data, but also in setting up experiments. One consideration is the choice of a sample size. In making interval estimates of population means, we have seen that each inference goes hand in hand with an associated confidence-level statement. Generally, if we want a smaller, more precise interval estimate, we either decrease the degree of confidence or increase the sample size. Similarly, if we want to increase the degree of confidence, we may either accept a wider interval estimate or increase the sample size. Choosing a larger sample size seems always desirable; in the real world, however, this decision involves time and cost considerations.*

KEY EXAMPLE

Ball bearings are manufactured by a process that results in a standard deviation in diameter of 0.025 inch. What sample size should be chosen if one wishes to be 99% sure of knowing the diameter to within ±0.01 inch?

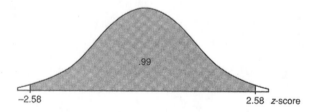

Answer: We have $\sigma_{\bar{x}} = \sigma / \sqrt{n} = 0.025 / \sqrt{n}$ and $2.58\sigma_{\bar{x}} \leq 0.01$. Thus $2.58(0.025/\sqrt{n}) \leq 0.01$. Algebraically we find that $\sqrt{n} \geq 2.58(0.025)/0.01 = 6.45$, so $n \geq 41.6$. We choose a sample size of 42.

A sociologist is designing an experiment to determine the mean age of U.S. citizens who have strong opinions against committing funds for military operations in Central America. She has determined that, for a 90% confidence estimate of the mean age within ±3.0 years, she will need to survey 100 individuals having the specified opinions. What would be the sample size for a 90% confidence estimate of the mean age within ±1.5 years? Within ±1.0 year?

Answer: We have $\sigma_{\bar{x}} = \sigma/\sqrt{100}$ and $1.645\sigma_{\bar{x}} \le 3.0$, or $1.645(\sigma/\sqrt{100}) \le 3.0$. We want to know n_1 and n_2, where $1.645(\sigma/\sqrt{n_1}) \le 1.5$ and $1.645(\sigma/\sqrt{n_2}) \le 1.0$. Algebraically, we find $\sqrt{n_1} \ge 1.645\sigma/1.5$, and $\sqrt{n_2} \ge 1.645\sigma/1.0$, and from the known sample size we have $1.645\sigma \le 3.0\sqrt{100}$. Thus $\sqrt{n_1} \ge 3.0\sqrt{100}/1.5$, so $n_1 \ge 400$ and $\sqrt{n_2} \ge 3.0\sqrt{100}/1.0$, so $n_2 \ge 900$.

There are two points worth noting. First, what would have been the result if, for example, 95% had been used instead of 90% for the confidence level?

Answer: Each 1.645 would have been replaced by 1.96, and instead of dividing 1.645 by 1.645 we would divide 1.96 by 1.96, but the resulting answers of 400 and 900 would not have changed.

Second, how much must the sample size be increased in order to cut the interval estimate in half? To a third?

Answer: To cut the interval estimate in half (from ±3.0 to ±1.5), we would have to increase the sample size fourfold (from 100 to 400). To cut the interval estimate to a third (from ±3.0 to ±1.0), we would have to increase the sample size ninefold (from 100 to 900).

More generally, if we want to divide the interval estimate by d without effecting the confidence level, we must increase the sample size by a multiple of d^2.

Key 53 Hypothesis test of the mean

OVERVIEW *Closely related to the problem of estimating a population mean is the problem of testing a hypothesis about a population mean. For example, a consumer protection agency might determine an interval estimate for the mean nicotine content of a particular brand of cigarettes, or, alternatively, it might test a manufacturer's claim about the mean nicotine content of his cigarettes. An agricultural researcher could find an interval estimate for the mean productivity gain caused by a specific fertilizer, or, alternatively, she might test the developer's claimed mean productivity gain. A social scientist might ascertain an interval estimate for the mean income level of migrant farmers, or, alternatively, he might test a farm bureau's claim about the mean income level.*

The general testing procedure is to choose a specific hypothesis to be tested, called the **null hypothesis,** pick an appropriate sample, and use measurements from the sample to determine the likelihood of the null hypothesis. Conclusions are never stated with absolute certainty, but rather with associated significance levels. There are two types of possible errors that we consider: the error of mistakenly rejecting a true null hypothesis and the error of mistakenly failing to reject a false null hypothesis.

KEY EXAMPLE

A manufacturer claims that the mean lifetime of an electronic component of his product is 1500 hours. A researcher believes that the true figure is lower and will test the 1500-hour claim by measuring the lifetime of each element in a sampling of components. The researcher decides that, if the sample average is less than 1450 hours, she will reject the manufacturer's claim. Alternatively, if the sample average is over 1450 hours, she will conclude that she does not have sufficient evidence to reject the 1500-hour claim.

The claim to be tested, the *null hypothesis,* labeled H_0, is usually what we want to disprove, and is stated in terms of a specific value for a population parameter. In this case

$$H_0: \mu = 1500$$

The *alternative hypothesis*, denoted as H_a, is usually what we want to establish, and is stated in terms of an inequality such as $<, >,$ or $\neq$. In this case

$$H_a: \mu < 1500$$

The testing procedure involves picking a sample and comparing the sample mean $\bar{x}$ to the claimed population mean μ. The strength of the sample statistic $\bar{x}$ can be gauged through its associated *P*-value, which is the probability of obtaining a sample statistic as extreme as the one obtained if the null hypothesis is assumed to be true. The smaller the *P*-value, the more significant the difference between the null hypothesis and the sample results.

Alternatively, we can decide on a *critical value c* to gauge the significance of the sample statistic. If the observed $\bar{x}$ is further from the claimed mean μ than is the critical value, then we say that there is sufficient evidence to reject the null hypothesis.

If the alternative hypothesis involves $<$ or $>$, then there is one *critical value c* that separates the null hypothesis *rejection region* from the *fail to reject* region. In this case

$$c = 1450 \quad \text{and} \quad \frac{\text{rejection region} \mid \text{fail to reject region}}{1450}$$

If the 1500-hour claim is true but the sample mean happens to be less than 1450, then the researcher will *mistakenly reject* the null hypothesis. This is called a **Type I error,** and the probability of committing such an error is called the α-**risk.** If the 1500-hour claim is false but the sample mean happens to be over 1450, then the researcher will *mistakenly fail to reject* the null hypothesis. This is called a **Type II error,** and the probability of committing such an error is called the β-**risk.**

Null Hypothesis ($\mu = 1500$)

		True	False
Sample Mean	Less than 1450	Type I error *mistakenly reject true claim*	Correct decision *reject false claim*
	More than 1450	Correct decision *do not reject true claim*	Type II error *mistakenly fail to reject false claim*

Key 54 Critical values, α-risks, and
P-values

OVERVIEW *We test a null hypothesis by picking a sample and comparing the sample mean to the claimed population mean. If the resulting P-value of the test is small enough, we say there is evidence to reject the null hypothesis. Alternatively, we can choose a critical value, or calculate a critical value given an acceptable α-risk, and note if the sample mean is further from the claimed mean than is the critical value. We assume we have simple random samples from approximately normally distributed populations.*

KEY EXAMPLE

A manufacturer claims that a new brand of air-conditioning unit uses only 6.5 kilowatts of electricity per day. A consumer agency believes the true figure is higher and runs a test on a sample of size 50. If the sample mean is 7.0 kilowatts with a standard deviation of 1.4, should the manufacturer's claim be rejected at a significance level of 5%? Of 1%?

Answer: H_0: $\mu = 6.5$, H_a: $\mu > 6.5$, and $\alpha = .05$ or $.01$.

Here $\sigma_{\bar{x}} = \sigma / \sqrt{n} \approx s / \sqrt{n} = 1.4 / \sqrt{50} = 0.198$. The *t*-score of 7.0 is $(7.0 - 6.5)/0.198 = 2.53$.

The critical *t*-values for the 5% and 1% tests are approximately 1.677 and 2.406, respectively, and we have both 2.53 > 1.677 and 2.53 > 2.406. [On the TI-83, STAT → TESTS → T-Test, then Inpt:Stats, μ0:6.5, $\bar{x}$:7, Sx:1.4, n:50, μ:>μ0, Calculate yields $t = 2.525$ and $P = .0074$.] The consumer agency should reject the manufacturer's claim at both the 5% and 1% significance levels. When $P < .01$, we usually say that there is *strong* evidence to reject H_0.

KEY EXAMPLE

A cigarette industry spokesperson remarks that current levels of tar are no more than 5 milligrams per cigarette. A reporter does a quick check on 15 cigarettes representing a cross section of the market. What conclusion is reached if the sample mean is 5.63 milligrams of tar with a standard deviation of 1.61? Assume a 10% significance level.

Answer:

$$H_0: \mu = 5, \quad H_a: \mu > 5, \quad \alpha = .10$$

$$\sigma_{\bar{x}} = \frac{s}{\sqrt{n}} = \frac{1.61}{\sqrt{15}} = 0.42$$

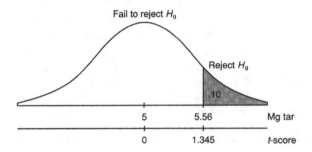

With df = 15 − 1 = 14, and $\alpha = .10$, the critical t-score is 1.345. The critical number of milligrams of tar is 5 + 1.345(0.42) = 5.56. Since 5.63 > 5.56, the industry spokesperson's remarks should be rejected at the 10% significance level.

What is the conclusion at the 5% significance level?

Answer: In this case, the critical t-score is 1.761; the critical tar level is 5 + 1.761(0.42) = 5.74; and since 5.63 < 5.74, the remarks cannot be rejected at the 5% significance level.

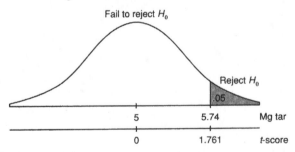

What is the *P*-value?

Answer: Without a software program we can't calculate the *P*-value exactly. However the *t*-score for 5.63 is (5.63 − 5)/0.42 = 1.5, and because 1.5 is between 1.345 and 1.761, we can conclude that *P* is between .05 and .10. [On the TI-83, T-Test gives *P* = .0759.]

KEY EXAMPLE

A local chamber of commerce claims that the mean sale price for homes in the city is \$90,000. A real estate salesperson notes eight recent sales of \$75,000, \$102,000, \$80,000, \$85,000, \$79,000, \$95,000, \$98,000, and \$62,000.

How strong is the evidence to reject the chamber of commerce claim?

Answer: H_0: $\mu = 90{,}000$ and H_a: $\mu \neq 90{,}000$. A calculator gives $\bar{x} = 84{,}500$ and $s = 13{,}341.7$, and then one calculates $\sigma_{\bar{x}} \approx s/\sqrt{n} = 13{,}341.7/\sqrt{8} = 4717$. The t-score of 84,500 is $(84{,}500 - 90{,}000)/4717 = 1.17$. With df $= 8 - 1 = 7$, we note that 1.17 is less than 1.415, which corresponds to a tail probability of .10. Doubling because this is a two-sided test, we see that the P-value is greater than .20. With such a high P, we conclude that there is no evidence to reject the chamber of commerce claim. [On the TI-83 put the data in a list and then use Data under T-Test to obtain $P = .2818$.]

Key 55 Type II errors and power

OVERVIEW *Why not always choose the α-risk to be extremely small, such as .001 or .0001, and so virtually eliminate the possibility of mistakenly rejecting a correct null hypothesis? The problem is that doing this would simultaneously increase the chance of never rejecting the null hypothesis, even if it were false. Thus we are led to a discussion of the Type II error, that is, a mistaken failure to reject a false null hypothesis. The probability of a Type II error is called the β-risk, and there is a different β-value for each possible correct value for the population parameter. For each β, 1–β is called the **power** of the test against the associated correct value. The **power** of a hypothesis test is the probability that a Type II error is not committed.*

An illustration of the difference between a Type I and a Type II error is as follows: Suppose the null hypothesis is that all systems are operating satisfactorily with regard to a NASA liftoff. A Type I error would be to delay the liftoff mistakenly, thinking that something was malfunctioning when everything was actually okay. A Type II error would be to fail to delay the liftoff mistakenly, thinking everything was okay when something was actually malfunctioning. The power is the probability of recognizing a particular malfunction. (Note the complementary aspect of power, a "good" thing, with Type II error, a "bad" thing.)

KEY EXAMPLE

City planners are trying to decide among various parking-plan options ranging from more on-street spaces to multilevel facilities to spread-out small lots. Before making a decision, they wish to test the downtown merchants' claim that shoppers park for an average of only 47 minutes in the downtown area. The planners have decided to tabulate parking durations for 225 shoppers and to reject the merchants' claim if the sample mean is over 50 minutes. What is the probability of a Type II error if the true value is 48? If the true value is 51? Assume that the standard deviation in parking durations is 27 minutes.

Answer: We have:

$$H_0: \mu = 47, \quad H_a: \mu > 47$$

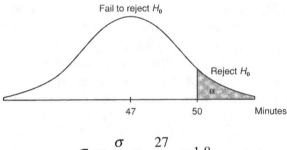

$$\sigma_{\bar{x}} = \frac{\sigma}{\sqrt{n}} = \frac{27}{\sqrt{225}} = 1.8$$

If the true mean parking duration is 48, then the normal curve should be centered at 48. In this case, the *z*-score of 50 is $(50 - 48)/1.8 = 1.11$. Using Table A, we calculate the β-risk (probability of failure to reject H_0) to be $.5000 + .3665 = .8665$. (The power is $1 - .8665 = .1335$.)

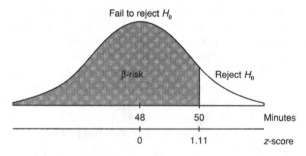

If the true mean parking duration is 51, then the normal curve should be centered at 51. The critical value is still 50, with a *z*-score now of $(50 - 51)/1.8 = -0.56$. We use Table A to find the β-risk to be $.5000 - .2123 = .2877$. (The power is $1 - .2877 = .7123$.)

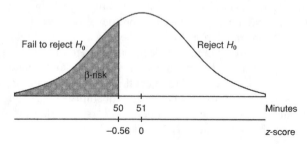

Note that the further the true value is (in the suspected direction) from the claimed mean, the smaller the probability is of failing to reject the false claim.

In many situations we start with a significance level, use this to calculate the critical value, and then find particular β-risks.

KEY EXAMPLE

A geologist claims that a particular rock formation will yield a mean of 24 pounds of a chemical per ton of excavation. A company, fearful that the true amount will be less, plans to run a test on a sample of 50 tons.

If the standard deviation between tons is 5.8 pounds, what is the critical value? Assume a 1% significance level. What is the probability of a Type II error if the true mean is 22? Is 20?

Answer: We have:

$$H_0: \mu = 24, \quad H_a: \mu < 24, \quad \alpha = .01$$

$$\sigma_{\bar{x}} = \frac{\sigma}{\sqrt{n}} = \frac{5.8}{\sqrt{50}} = 0.82$$

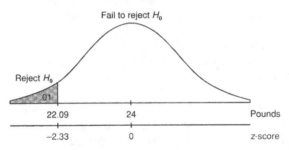

With $\alpha = .01$ the critical z-score is -2.33, and the critical poundage value is $24 - 2.33(0.82) = 22.09$. Thus, the decision rule is to reject H_0 if the sample mean is less than 22.09, and fail to reject if it is more than 22.09.

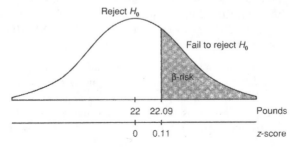

If the true mean is 22 pounds per ton of rock, then the z-score of 22.09 is $(22.09 - 22)/0.82 = 0.11$, and so the β-risk is $.5000 - .0438 = .4562$. (The power is $1 - .4562 = .5438$.)

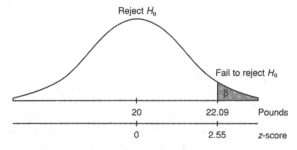

Reject H_0

Fail to reject H_0

β

| 20 | 22.09 | Pounds |

| 0 | 2.55 | z-score |

If the true mean is 20 pounds per ton of rock, then the z-score of 22.09 is $(22.09 - 20)/0.82 = 2.55$, and so the β-risk is $.5000 - .4946 = .0054$. (The power is $1 - .0054 = .9946$.)

KEY EXAMPLE

A factory manager claims that the plant's smokestacks spew forth only 350 pounds of pollution per day. A government investigator suspects that the true value is higher and plans a hypothesis test with a critical value of 375 pounds. Suppose the standard deviation in daily pollution poundage is 150. What is the α-risk for a sample size of 100 days? 200 days? What are the associated β-risks if the true mean is 385 pounds?

Answer: We have: H_0: $\mu = 350$, H_a: $\mu > 350$, and $c = 375$

With $n = 100$, $\sigma_{\bar{x}} = 150/\sqrt{100} = 15.0$, the z-score of 375 is $(375–350)/15.0 = 1.67$, and the corresponding α-risk from Table A is $.5000 - .4525 = .0475$. With $n = 200$, $\sigma_{\bar{x}} = 150/\sqrt{200} = 10.6$, the z-score of 375 is $(375 - 350)/10.6 = 2.36$, and the corresponding α-risk is $.5000 - .4909 = .0091$.

With $n = 100$ and a mean of 358, the z-score of 375 is $(375 - 385)/15.0 = -0.67$ with a corresponding β-risk of $.5000 - .2486 = .2514$. With $n = 200$ and a mean of 385, the z-score of 375 is $(375 - 385)/10.6 = -0.94$ with a corresponding β-risk of $.5000 - .3264 = .1736$.

We note that in this particular example *increasing* the sample size (from 100 to 200) resulted in *decreases* in both the α-risk (from .0475 to .0091) and the β-risk (from .2514 to .1736). More generally, if α is held fixed and the sample size is increased, then the β-risk will decrease. Furthermore, by increasing the sample size and adjusting the critical value, it is always possible to decrease both the α-risk and the β-risk. Increasing the sample size or increasing the significance level are both ways of increasing the power.

Key 56 Confidence interval estimate for the difference of two means

OVERVIEW *Many real-life applications of statistics involve comparisons of two populations. For example, is the average weight of laboratory rabbits receiving a special diet greater than that of rabbits on a standard diet? Which of two accounting firms gives a higher mean starting salary? Is the life expectancy of a coal miner less than that of a school-teacher? To compare the means of two populations, we compare the means of two samples, one from each population.*

In many situations it is clear that the mean of one population is higher than that of another, and we would like to estimate the difference. Using samples, we cannot find this difference *exactly,* but we will be able to say with a certain confidence that the difference lies in a certain *interval.* We follow the same procedure as set forth for a single mean, this time using $\mu_1 - \mu_2$, $\sqrt{\sigma_1^2 / n_1 + \sigma_2^2 / n_2}$, and $\bar{x}_1 - \bar{x}_2$ in place of μ, $\sigma / \sqrt{n}$, and $\bar{x}$, respectively. [Remark: the formula for standard deviation is derived from the fact that the variance of a set of differences is equal to the sum of the variances of the individual sets.]

In making calculations and drawing conclusions from specific samples, it is important both that the samples be *simple random samples* and that they be taken *independently* of each other. We also assume that the two parent populations are approximately normally distributed.

KEY EXAMPLE

A 30-month study is conducted to determine the difference in the rates of accidents per month between two departments in an assembly plant. Suppose the first department averaged 12.3 accidents per month with a standard deviation of 3.5, while the second averaged 7.6 accidents with a standard deviation of 3.4. Determine a 95% confidence interval estimate for the difference in accidents per month.

Answer:

$$n_1 = 30 \quad n_2 = 30$$
$$\bar{x}_1 = 12.3 \quad \bar{x}_2 = 7.6$$
$$s_1 = 3.5 \quad s_2 = 3.4$$

$$\sigma_{\bar{x}_1 - \bar{x}_2} = \sqrt{\frac{\sigma_1^2}{n_1} + \frac{\sigma_2^2}{n_2}} \approx \sqrt{\frac{s_1^2}{n_1} + \frac{s_2^2}{n_2}} = \sqrt{\frac{(3.5)^2}{30} + \frac{(3.4)^2}{30}} = 0.89$$

The observed difference is $12.3 - 7.6 = 4.7$, and with df $= (n_1 - 1) + (n_2 - 1) = 58$, the critical *t*-scores are ± 2.00. Thus the confidence interval estimate is $4.7 \pm 2.00(0.89) = 4.7 \pm 1.78$. We are 95% confident that the first department has between 2.92 and 6.48 more accidents per month than the second department. [On the TI-83, 2-SampTInt gives (2.9167, 6.4833).]

KEY EXAMPLE

A survey is run to determine the difference in the cost of groceries in suburban stores versus inner-city stores. A preselected group of items is purchased in a sample of 45 suburban and 35 inner-city stores, and the following data are obtained:

Suburban stores	Inner-city stores
$n_1 = 45$	$n_2 = 35$
$\bar{x}_1 = \$36.52$	$\bar{x}_2 = \$39.40$
$s_1 = \$1.10$	$s_2 = \$1.23$

Find a 90% confidence interval estimate for the difference.

Answer:

$$\sigma_{\bar{x}_1 - \bar{x}_2} = \sqrt{\frac{(1.10)^2}{45} + \frac{(1.23)^2}{35}} = 0.265$$

The observed difference is $36.52 - 39.40 = -2.88$, the critical *t*-scores are ± 1.664, and the confidence interval estimate is $-2.88 \pm 1.664(0.265) = -2.88 \pm 0.44$. Thus we are 90% confident that the selected group of items costs between $2.44 and $3.32 *less* in suburban stores than in inner-city stores. [On the TI-83, 2-SampTInt gives (−3.321, −2.439).]

A common sample size can be chosen by a method similar to that used for the one-sample case.

KEY EXAMPLE

A hardware store owner wishes to determine the difference in drying times between two brands of paints. Suppose the standard deviation between cans in each population is 2.5 minutes. How large a sample (same number) of each must be used if the owner wishes to be 98% sure of knowing the difference to within 1 minute?

Answer:

$$\sigma_{\bar{x}_1 - \bar{x}_2} = \sqrt{\frac{(2.5)^2}{n} + \frac{(2.5)^2}{n}} = \frac{3.536}{\sqrt{n}}$$

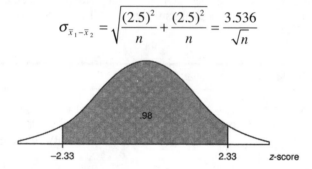

With a critical z-score of 2.33, we have $2.33(3.536/\sqrt{n}) \leq 1$, so $\sqrt{n} \geq 8.24$ and $n \geq 67.9$. Thus the owner should test samples of 68 paint patches from each brand.

Key 57 Hypothesis test for the difference of two means

OVERVIEW *The fact that a sample mean from one population is greater than a sample mean from a second population does not automatically lead to a similar conclusion about the means of the populations themselves. Two points need to be stressed. First, means of samples from the same population vary from each other. Second, what we are really comparing are confidence interval estimates, not just single points.*

In this situation the null hypothesis is usually that the means of the populations are the same, or, equivalently, that their difference is 0:

$$H_0: \mu_1 - \mu_2 = 0$$

The alternative hypothesis is then:

$$H_a: \mu_1 - \mu_2 < 0, \quad H_a: \mu_1 - \mu_2 > 0, \quad \text{or} \quad H_a: \mu_1 - \mu_2 \neq 0$$

The first two possibilities lead to one-sided (one-tailed) tests, while the third possibility leads to two-sided (two-tailed) tests.

KEY EXAMPLE

A salesman believes that his company's computer has more average downtime per week than does a similar computer manufactured by a competitor. Before bringing his concern to his director, the salesman gathers data and runs a hypothesis test. In a sample of 40 week-long periods in different firms using his company's product, the average downtime was 125 minutes with a standard deviation of 37 minutes. However, 35 week-long periods involving the competitor's computer yield an average downtime of only 115 minutes with a standard deviation of 43 minutes. What conclusion should be drawn, assuming a 10% significance level?

Answer:

$$H_0: \mu_1 - \mu_2 = 0, \quad H_a: \mu_1 - \mu_2 > 0, \quad \alpha = .10$$

$$\sigma_{\bar{x}_1 - \bar{x}_2} = \sqrt{\frac{(37)^2}{40} + \frac{(43)^2}{35}} = 9.33$$

The difference in sample means is $125 - 110 = 10$, and the t-score is $10/9.33 = 1.07$. Using Table B, we find the P-value to be over .10. Thus, the observed difference is not significant at the 10% significance level. The sales representative does not have sufficient evidence that his company's computer has more downtime. [On the TI-83, STAT $\rightarrow$ TESTS $\rightarrow$ 2-SampTTest, then Inpt:Stats, $\bar{x}$1:125, Sx1:37, n1:40, $\bar{x}$2:115, Sx2:43, n2:35, μ1:>μ2 and Calculate yields $t = 1.0718$ and $P = .1438$.]

KEY EXAMPLE

A store manager wishes to determine whether there is a significant difference between two trucking firms with regard to the handling of egg cartons. In a sample of 200 cartons on one firm's truck, there was an average of 0.7 broken egg per carton with a standard deviation of 0.31, while a sample of 300 cartons on the second firm's truck showed an average of 0.775 broken egg per carton with a standard deviation of 0.42. Is the difference between the averages significant at a significance level of 5%? At a level of 1%?

Answer:

$$H_0: \mu_1 - \mu_2 = 0, \quad H_a: \mu_1 - \mu_2 \neq 0, \quad \alpha = .05$$

$$\sigma_{\bar{x}_1 - \bar{x}_2} = \sqrt{\frac{(0.31)^2}{200} + \frac{(0.42)^2}{300}} = 0.0327$$

Because there is no mention that it is believed that one firm does better than the other, this is a two-sided test. The observed difference of $0.7 - 0.775 = -0.075$ has a t-score of $-0.075/0.0327 = -2.29$. Using Table B, we find that this gives a probability between .01 and .025 (closer to .01). The test is two-sided, and so the P-value is between .02 and .05 (closer to .02). So the observed difference is statistically significant at the 5% level but not at the 1% level. With $.01 < P < .05$, we usually say that there is *moderate* evidence to reject H_0. [On the TI-83, 2-SampTTest gives $P = .0222$.]

Key 58 Theme exercises with answers

OVERVIEW *Sample questions of the type that might appear on homework assignments and tests are presented with answers.*

- In a certain plant, batteries are being produced with life expectancies that have a variance of 5.76 months². Suppose the mean life expectancy in a sample of 64 batteries is 12.35 months. Find a 90% confidence interval estimate of life expectancy for all batteries produced in this plant.

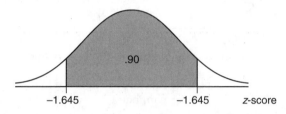

Answer: The standard deviation of the population is $\sigma = \sqrt{5.76} = 2.4$, and the standard deviation of the sample means is $\sigma_{\bar{x}} = \sigma / \sqrt{n} = 2.4 / \sqrt{64} = 0.3$. The 90% confidence interval estimate for the population mean is $12.35 \pm 1.645(0.3) = 12.35 \pm 0.4935$. Thus we are 90% certain that the mean life expectancy of the batteries is between 11.8565 and 12.8435 months.

What would be the 90% confidence interval estimate if the sample mean of 12.35 had come from a sample of 100 batteries?

Answer: The standard deviation of sample means would then have been $\sigma_{\bar{x}} = \sigma / \sqrt{n} = 2.4 / \sqrt{100} = 0.24$, and the 90% confidence interval estimate would be $12.35 \pm 1.645(0.24) = 12.35 \pm 0.3948$.

Note that, when the sample size increased (from 64 to 100), the same sample mean resulted in a narrower, more specific interval (±0.3948 instead of ±0.4935).

- A government investigator plans to test for the mean quantity of a particular pollutant that a manufacturer is dumping per day into a river. She needs an estimate that is within 50 grams at the 95% confidence level. If previous measurements indicate that the variance is approximately 21,800 grams², how many days should she include in the sample?

Answer:

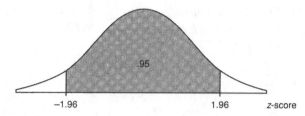

We have $\sigma^2 = 21,800$, so $\sigma = \sqrt{21,800} = 147.65$. Then $\sigma_{\bar{x}} = \sigma/\sqrt{n} = 147.65/\sqrt{n}$, and $1.96\sigma_{\bar{x}} \leq 50$. Thus $1.96(147.65/\sqrt{n}) \leq 50$, $\sqrt{n} \geq 1.96(147.65)/50 = 5.788$, and $n \geq 33.5$. Therefore, the investigator should sample 34 days' output.

- A coffee-dispensing machine is supposed to drop 8 ounces of liquid into each paper cup, but a consumer believes that the actual amount is less. As a test he plans to obtain a sample of 36 cups of the dispensed liquid, and, if the mean content is less than 7.75 ounces, to reject the 8-ounce claim. If the machine operates with a standard deviation of 0.9 ounce, what is the α-risk?

Answer: We have:

$$H_0: \mu = 8, \quad H_a: \quad \mu < 8$$

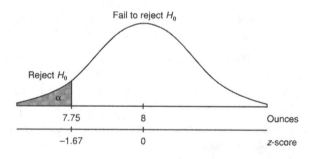

The standard deviation of sample means is $\sigma_{\bar{x}} = \sigma/\sqrt{n} = 0.9/\sqrt{36} = 0.15$. The z-score of 7.5 is $(7.75 - 8)/0.15 = -1.67$. Using Table A, we obtain $\alpha = .5000 - .4525 = .0475$. Thus, if the 8-ounce claim is correct, there is a .0475 probability that the consumer will still obtain a sample mean less than 7.75 and mistakenly reject the claim.

- A pharmaceutical company claims that a medication will produce a desired effect for a mean of 58.4 minutes. A government researcher runs a hypothesis test of 250 patients and tabulates the following data with reference to the durations of the effect in minutes: $\Sigma x = 14{,}875$ and $\Sigma(x - \bar{x})^2 = 17{,}155$. Should the company's claim be rejected at a significance level of 10%? Of 2%?

Answer:

$$\bar{x} = \frac{\Sigma x}{n} = \frac{14{,}875}{250} = 59.5, \quad s = \sqrt{\frac{\Sigma(x - \bar{x})^2}{n-1}} = \sqrt{\frac{17{,}155}{249}} = 8.3$$

and

$$\sigma_{\bar{x}} \approx \frac{s}{\sqrt{n}} = \frac{8.3}{\sqrt{250}} = 0.525$$

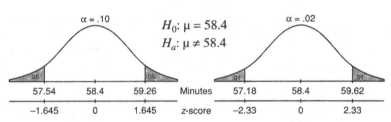

$H_0: \mu = 58.4$
$H_a: \mu \neq 58.4$

For $\alpha = .10$, the critical values in minutes are $58.4 \pm 1.645(0.525)$, or 57.54 and 59.26. For $\alpha = .02$, the critical values in minutes are $58.4 \pm 2.33(0.525)$, or 57.18 and 59.62. Now $59.5 > 59.26$, but $59.5 < 59.62$, so the researcher would reject the company's claim at the 10% significance level, but not at the 2% significance level. In other words, if she is willing to be wrong 10 times in 100, she would reject the company's claim; but if she is willing to accept only 2 errors in 100 such decisions, she would conclude that the evidence is not strong enough to reject the claim.

Again, the *P*-value can be used to specifically measure the strength of evidence for rejection of H_0. In this example, the *z*-score of 59.5 is $(59.5 - 58.4)/0.525 = 2.10$. Using Table A, we find the corresponding probability to be $.5000 - .4821 = .0179$. Doubling because the test is two-sided results in a *P*-value of $2(.0179) = .0358$.

- A medical research team claims that high vitamin C intake increases endurance. In particular, 1000 mg of vitamin C per day for a month should add an average of 4.3 minutes to the possible length of maximum physical effort. Army training officers believe the claim is exaggerated and plan a test on 400 soldiers. If the standard deviation of added minutes is 3.2, find the critical values for a significance level of 5% and then of 1%. In each case find also the probability of a Type II error if the true mean increase is only 4.2 minutes.

Answer: We have:

$$H_0 : \mu = 4.3, \quad H_a : \mu < 4.3$$

$$\sigma_{\bar{x}} = \frac{3.2}{\sqrt{400}} = 0.16$$

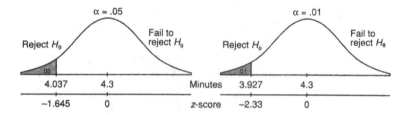

Corresponding to a z-score of -1.645 is $4.3 - 1.645(0.16) = 4.037$; corresponding to a z-score of -2.33 is $4.3 - 2.33(0.16) = 3.927$.

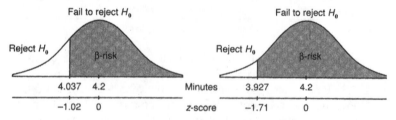

With a mean of 4.2, the z-score of 4.038 is $(4.038 - 4.2)/0.16 = -1.02$, and the β-risk is $.5000 + .3461 = .8461$. With a mean of 4.2, the z-score of 3.927 is $(3.927 - 4.2)/0.16 = -1.71$, and the β-risk is $.5000 + .4594 = .9564$.

We note that *decreasing* the α-risk (from .05 to .01) led to an *increase* in the β-risk (from .8438 to .9564). This is always the case if the sample size is held constant.

- A trucking firm conducts a test to compare the life expectancies of two brands of tires. Shipments of 1000 and 1500 tires, respectively, are received and marked, and the truck mileages are noted when the tires are replaced. The resulting raw data are as follows:

Brand F	Brand G
$n_1 = 1000$	$n_2 = 1500$
$\Sigma x = 22{,}350{,}000$	$\Sigma x = 36{,}187{,}500$
$\Sigma(x - \bar{x})^2 = 9{,}600{,}000{,}000$	$\Sigma(x - \bar{x})^2 = 15{,}800{,}000{,}000$

Determine a 99% confidence interval estimate for the difference in life expectancies.

Answer: Calculations of the means and standard deviations yield:

Brand F	Brand G
$\bar{x}_1 = \dfrac{22{,}350{,}000}{1000} = 22{,}350$	$\bar{x}_2 = \dfrac{36{,}187{,}500}{1500} = 24{,}125$
$s_1 = \sqrt{\dfrac{9{,}600{,}000{,}000}{1000 - 1}} = 3100$	$s_2 = \sqrt{\dfrac{15{,}800{,}000{,}000}{1500 - 1}} = 3246.6$

$$\sigma/\bar{x}_1 - \bar{x}_2 = \sqrt{\frac{(3100)^2}{1000} + \frac{(3246.6)^2}{1500}} = 129.0$$

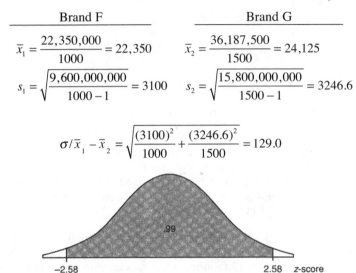

The observed difference is $22{,}350 - 24{,}125 = -1775$, and the critical z-scores are ±2.58. Since $-1775 \pm 2.58(129) = -1775 \pm 333$, the trucking firm can be 99% sure that brand F tires average between 1442 and 2108 miles less in life expectancy than brand G tires.

- In a sleeping laboratory experiment, 16 volunteers sleep an average of 7.4 hours with a standard deviation of 1.3 hours. What would be a 90% confidence interval estimate for the mean number of hours that people sleep at night?

Answer: The standard deviation of the sample means $\sigma_{\bar{x}} = 1.3/\sqrt{16} = 0.325$. With $16 - 1 = 15$ degrees of freedom, and 5% in each tail, the appropriate t-scores are ±1.753. Thus we can be 90% certain that the mean number of hours that people sleep at night is in the range $7.4 \pm 1.753(0.325) = 7.4 \pm 0.57$, or between 6.83 and 7.97 hours. [On the TI-83, TInterval gives (6.8303, 7.9697).]

- A broker notes that the percentage gains in per-share value for stock in eight leisure-time companies during a particular 1-year period were 8.2, 9.5, 4.2, 10.0, 6.7, 6.6, 9.3, and 7.9. Find a 95% confidence interval estimate for the percentage gain in per-share value for leisure-time company stocks.

Answer: The sample mean is

$$\bar{x} = \frac{8.2 + 9.5 + 4.2 + 10.0 + 6.7 + 6.6 + 9.3 + 7.9}{8} = 7.8$$

and the standard deviation is

$$s = \sqrt{\frac{0.4^2 + 1.7^2 + 3.6^2 + 2.2^2 + 1.1^2 + 1.2^2 + 1.5^2 + 0.1^2}{7}} = 1.918$$

Also,

$$\sigma_{\bar{x}} = \frac{1.918}{\sqrt{8}} = 0.678$$

With df = 8 − 1 = 7, and 2.5% in each tail, the *t*-scores are ±2.365. A 95% confidence interval estimate is given by 7.8 ± 2.365 (0.678) = 7.8 ± 1.6, or between 6.2 and 9.4 percent. [On the TI-83, putting the data in a list and using Data under TInterval gives (6.1962, 9.4038).]

- An IRS spokesperson claims that the average deduction for medical care is $1250. A taxpayer who believes that the real figure is lower samples 12 families and comes up with a mean of $934 and a standard deviation of $616. What conclusion should the taxpayer reach at a 5% significance level?

Answer:

$$H_0 : \mu = 1250, \quad H_a : \mu < 1250, \quad \alpha = .05$$

$$\sigma_{\bar{x}} = \frac{616}{\sqrt{12}} = 177.8$$

With df = 12 − 1 = 11, and α = .05, the critical *t*-score is 1.796. The critical number of dollars is 1250 − 1.796(177.8) = 930.7. Since 934 > 930.7, at the 5% significance level the taxpayer should not reject the IRS spokesperson's claim. [On the TI-83, T-Test gives a *P*-value of .0516, which is > .05.]

- The weight of an aspirin tablet is 300 mg according to the bottle label. Should an FDA investigator reject the label if she weighs seven tablets and obtains weights of 299, 300, 305, 302, 299, 301, and 303 mg? Use a 2% significance level for this two-tailed test.

Answer:

$$H_0 : \mu = 300, \quad H_a : \mu \neq 300, \quad \alpha = .02$$

$$\bar{x} = \frac{2109}{7} = 301.3$$

$$s = \sqrt{\frac{29.43}{6}} = 2.21$$

$$\sigma_{\bar{x}} = \frac{2.21}{\sqrt{7}} = 0.835$$

For $\alpha = .02$, we place a probability of .01 on each side, and note that df $= 7 - 1 = 6$ gives critical t-scores of ± 3.143. Critical weights are $300 \pm 3.143(0.835) = 300 \pm 2.62$, or 297.38 and 302.62 mg. Since 301.3 is in this range, at the 2% significance level the FDA investigator should not reject the label. [On the TI-83, putting the data in a list and using Data under T-Test gives $P = .1755$.]

Theme 8   CHI-SQUARE ANALYSIS

*I*n this theme we consider three types of problems involving similar analyses. The first concerns whether or not some observed distributional outcome fits some previously specified pattern. Perfect fits very rarely exist, and we must develop a procedure for measuring the significance of a loose fit. The second type of problem concerns whether two variables are independent or have some relationship. Here we use the same kind of measurement as above to judge the significance of a loose fit with the theoretical pattern based on independence. The third type of problem concerns a single variable with a comparison between samples from two or more populations. The kinds and the variety of problems we will be considering are indicated by the following examples.

A geneticist might test whether inherited traits can be explained by a *binomial distribution*. A physicist might look to a *Poisson distribution* to further understand alpha-particle emissions. A psychologist might study human intelligence patterns in terms of a *normal distribution*. In these and many other applications, the question arises as to how well observed data fit the pattern expected from some specified distribution.

A pollster might look at whether or not voter support for a candidate is *independent* of the voters' ethnic backgrounds. An efficiency expert might look at whether or not the likelihood of an accident is *independent* of which shift an employee works on. A psychologist might test whether or not certain mental abnormalities are *independent* of socioeconomic background.

A teachers' union might be interested in whether students, teachers, and administrators show the same distribution in types of cars driven. A guidance counselor randomly samples each class to see if the proportion of students excited by mathematics is different between freshmen, sophomores, juniors, and seniors.

As can be seen, the procedure described in the following keys has a wide range of applications. **Chi-square analysis** is one of the most useful statistical techniques, both where other tests are not applicable, and where other tests may be unnecessarily complicated.

Key 59 Chi-square calculation

OVERVIEW   *A critical question in statistics is whether or not some observed pattern of data fits some given distribution. Since a perfect fit cannot be expected, we must be able to look at discrepancies and make a judgment as to the "goodness of fit." The best information is obtained by squaring the discrepancy values and then appropriately **weighting** each difference. Such weighting is accomplished by dividing each difference by the expected value. The sum of these weighted differences or discrepancies is called **chi-square**, and is denoted as χ^2 (χ is the lowercase Greek letter chi):*

$$\chi^2 = \Sigma \frac{(\text{obs.} - \text{exp.})^2}{\text{exp.}}$$

KEY EXAMPLE

Census figures show that a city has 64% white, 25% black, and 11% Hispanic residents. A sampling of 350 new city employees shows 243 white, 80 black, and 27 Hispanic. Is the city hiring in the same racial pattern as its residents?

Answer: If the answer to the question is "yes," then among the 350 new employees one would expect approximately .64(350) = 224 white, .25(350) = 87.5 black, and .11(350)= 38.5 Hispanic. Chi-square gives a measurement of the differences between these expected numbers and the actual results.

$$\chi^2 = \frac{(243 - 224)^2}{224} + \frac{(80 - 87.5)^2}{87.5} + \frac{(27 - 38.5)^2}{38.5} = 5.69$$

The smaller the resulting χ^2-value, the better the fit. To decide how large a calculated χ^2-value must be to be significant, that is, to choose a critical value, we must understand how χ^2-values are distributed. A χ^2-distribution is not symmetrical and is always skewed to the right. Just as was the case with the *t*-distribution, there are distinct χ^2-distributions each with an associated df value (number of degrees of freedom). The larger the df value, the closer the χ^2-distribution is to a normal distribution.

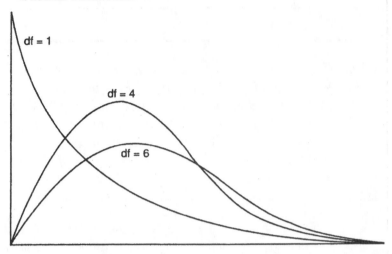

A large value of χ^2 may or may not be significant—the answer depends on which χ^2-distribution we are using. As with the t-distribution, Table C in the Appendix simply gives critical χ^2-values for the more commonly used percentages or probabilities.

There is a relationship between the χ^2-distribution and the normal distribution. While we do not give the mathematical explanation here, note that squaring our often used z-scores 1.645, 1.96, and 2.576, results in 2.706, 3.841, and 6.635, which are entries found in the first row of the χ^2-distribution table.

Key 60 Goodness-of-fit test for uniform distribution

OVERVIEW *Are muggings evenly distributed during the week? Do partners in a law firm bring in equal numbers of new clients? Do different brands of chemical fertilizers result in similar crop yields? In many examples the interest is in whether a **uniform** distribution is present. A hypothesis test is run on a sample, and chi-square is used to judge the significance of how far the observations differ from an even distribution.*

Our approach is similar to that developed earlier. There is the null hypothesis of a good fit, that is, that a uniform distribution correctly describes the situation, problem, or activity under consideration. Our observed data consist of one possible sample from a whole universe of possible samples. We ask about the chance of obtaining a sample with our observed discrepancies if the null hypothesis is really true. Finally, if the chance is too small, we reject the null hypothesis and say that the fit is not a good one.

KEY EXAMPLE

A grocery store manager wishes to determine whether a certain product will sell equally well in any of five locations in the store. Five displays are set up, one in each location, and the resulting numbers of the product sold are noted.

	Location				
	1	2	3	4	5
Number sold	43	29	52	34	48

Is there enough evidence that location makes a difference? Test at both the 5% and 10% significance levels.

Answer: We note that a total of 43 + 29 + 52 + 34 + 48 = 206 units were sold. If location doesn't matter, we would expect 206/5 = 41.2 units sold per location (uniform distribution).

<table>
<tr><td></td><td colspan="5">Location</td></tr>
<tr><td></td><td>1</td><td>2</td><td>3</td><td>4</td><td>5</td></tr>
<tr><td>Expected number</td><td>41.2</td><td>41.2</td><td>41.2</td><td>41.2</td><td>41.2</td></tr>
</table>

Thus

$$\chi^2 = \frac{(43-41.2)^2}{41.2} + \frac{(29-41.2)^2}{41.2} + \frac{(52-41.2)^2}{41.2} + \frac{(34-41.2)^2}{41.2}$$
$$+ \frac{(48-41.2)^2}{41.2} = 8.903$$

The degrees of freedom are the number of classes minus 1, that is, df = 5 − 1 = 4.

H_0: good fit for uniform distribution (product sells equally well in all 5 locations)

H_a: not a uniform distribution (product does not sell equally well in all 5 locations)

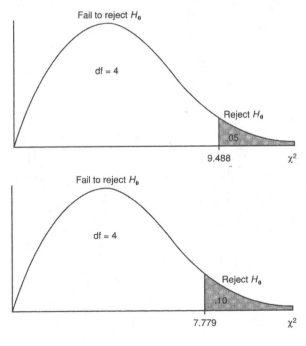

At the 5% level, with df = 4, the critical χ^2-value is 9.488, while at the 10% level the critical value is 7.779. Since 8.903 < 9.488 but 8.903 > 7.779, there is sufficient evidence to reject H_0 at the 10% level, but not at the 5% level. If the grocery store manager is willing to accept a 10% chance of committing a Type I error, then there is enough evidence to claim that location makes a difference. [With the TI-83, we can calculate the P-value: $P = \chi^2\text{cdf}(8.903,1000,4) = .0636$.]

Key 61 Goodness-of-fit test for prior distribution

OVERVIEW *Have the comparative percentages of college graduates who plan to become doctors, lawyers, and corporate executives changed over the percentages of 10 years ago? Do different species of fruit flies appear in the same ratios as they did before exposure to a particular chemical? Have the percentages of voters registered as Democrat, Republican, and Independent changed since before the last presidential election? To test such hypotheses we change the prior percentages into "expected" numbers by multiplying by the sample size, and then run a chi-square analysis.*

KEY EXAMPLE

Last year, at the 6 P.M. time slot, television channels 2, 3, 4, and 5 captured the entire audience with 30%, 25%, 20%, and 25%, respectively. During the first week of the new season, 500 viewers are interviewed. If viewer preference hasn't changed, what number is expected to watch each channel?

Answer:

$.30(500) = 150, \quad .25(500) = 125, \quad .20(500) = 100, \quad .25(500) = 125$

so we have

	Channel			
	2	3	4	5
Expected number	150	125	100	125

Suppose that the actual observed numbers are as follows:

	Channel			
	2	3	4	5
Observed number	139	138	112	111

Do these numbers indicate a change? Are the differences significant? We calculate:

$$\chi^2 = \Sigma \frac{(\text{obs.} - \text{exp.})^2}{\text{exp.}}$$

$$= \frac{(139 - 150)^2}{150} + \frac{(138 - 125)^2}{125} + \frac{(112 - 100)^2}{100} + \frac{(111 - 125)^2}{125}$$

$$= 5.167$$

Is 5.167 large enough for us to reject the null hypothesis of a good fit between the observed and the expected? To use the χ^2-table (Table C) in the Appendix, we must decide upon an α-risk and calculate the df value. The number of degrees of freedom for this problem is $4 - 1 = 3$. [Note that, while the observed values of 139, 138, and 112 can be freely chosen, the fourth value of 111 is absolutely determined because the total must be 500; thus df = 3.]

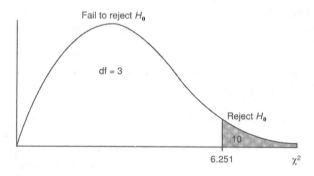

Thus, for example, at a 10% significance level, with df = 3, the critical χ^2 is 6.251. Since $5.167 < 6.251$, there is *not* sufficient evidence to reject H_0.

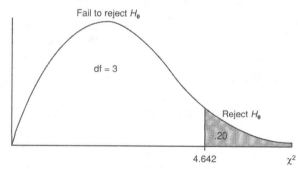

However, with df = 3 and a 20% significance level, the critical χ^2 is 4.642. Since 5.167 > 4.642, there *is* sufficient evidence to reject H_0. At the 20% level we would conclude that the fit is not good, and that viewer preference has changed.

Note: We can use the TI-83 to calculate the above χ^2 by putting the observed values in list L1, the expected values in L2, then $(L1-L2)^2/L2 \to L3$ and $\chi^2 = $ sum(L3) where "sum" is found under LIST $\to$ MATH. We can also use the TI-83 to calculate the *P*-value: χ^2cdf(5.167,1000,3) = .1600.

KEY EXAMPLE

A geneticist claims that four species of fruit flies should appear in the ratio 1:3:3:9. Suppose that a sample of 4000 flies contained 226, 764, 733, and 2277 flies of each species, respectively. At the 10% significance level, is there sufficient evidence to reject the geneticist's claim?

Answer: Since $1 + 3 + 3 + 9 = 16$, according to the geneticist the expected number of fruit flies of each species is as follows:

$$\frac{1}{16}(4000) = 250, \quad \frac{3}{16}(4000) = 750, \quad \frac{3}{16}(4000) = 750,$$
$$\frac{9}{16}(4000) = 2250$$

We calculate chi-square:

$$\chi^2 = \frac{(226 - 250)^2}{250} + \frac{(764 - 750)^2}{750} + \frac{(733 - 750)^2}{750}$$
$$+ \frac{(2277 - 2250)^2}{2250} = 3.27$$

With $4 - 1 = 3$ degrees of freedom and $\alpha = .10$, the critical χ^2 value is 6.251. Since 3.27 < 6.251, there is *not* sufficient evidence to reject H_0. At the 10% significance level, the geneticist's claim should not be rejected. [On the TI-83, $P = \chi^2$cdf(3.27,1000,3) = .3518.]

Key 62 Goodness-of-fit test for standard probability distributions

OVERVIEW *Chi-square is used to test whether an observed distribution fits the binomial, the Poisson, or the normal distribution.*

KEY EXAMPLE

Suppose that a commercial is run once on television, once on the radio, and once in a newspaper. The advertising agency believes that any potential consumer has a 20% chance of seeing the ad on TV, a 20% chance of hearing it on the radio, and a 20% chance of reading it in the paper. In a telephone survey of 800 consumers, the numbers claiming to have been exposed to the ad 0, 1, 2, or 3 times as follows:

	0	1	2	3
Observed number of people	434	329	35	2

At the 1% significance level, test the null hypothesis that the number of times any consumer saw the ad follows a binomial distribution with $p = 0.2$.

Answer: The complete binomial distribution with $p = .2$ and $n = 3$ is as follows:

$$P(0) = (.8)^3 = .512$$
$$P(1) = 3(.2)(.8)^2 = .384$$
$$P(2) = 3(.2)^2(.8) = .096$$
$$P(3) = (.2)^3 = .008$$

Multiplying each of these probabilities by 800 gives the expected number of occurrences:

$$.512(800) = 409.6, \quad .384(800) = 307.2, \quad .096(800) = 76.8,$$
$$.008(800) = 6.4$$

	0	1	2	3
Expected number of people	409.6	307.2	76.8	6.4

$$\chi^2 = \frac{(434 - 409.6)^2}{409.6} + \frac{(329 - 307.2)^2}{307.2} + \frac{(35 - 76.8)^2}{76.8} + \frac{(2 - 6.4)^2}{6.4}$$
$$= 28.776$$

H_0: good fit to a binomial distribution with $\pi = .2$

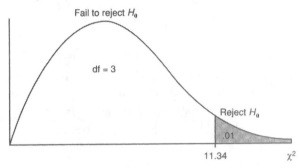

With df = 4 − 1 = 3 and $\alpha = .01$, the critical χ^2-value is 11.34. Since 28.776 > 11.34, there *is* sufficient evidence to reject H_0 and to conclude that the number of ads seen by each consumer does *not* follow a binomial distribution with $\pi = .2$. [On the TI-83, $P = \chi^2\text{cdf}(28.776,1000,3) = .0000025$.]

KEY EXAMPLE

For a hospital study of 450 patients with a particular form of cancer, the following chart shows the number of these patients who survived each of the given numbers of years.

	Survival Years				
	0	1	2	3	4 or more
Observed number of patients	60	110	125	88	67

Test the null hypothesis that the distribution follows a Poisson distribution with $\mu = 2.1$. Assume a 2.5% significance level.

Answer: The Poisson probabilities are:

$$P(0) = \quad e^{-2.1} \quad\quad = .122$$
$$P(1) = \quad 2.1e^{-2.1} \quad = .257$$
$$P(2) = \frac{(2.1)^2}{2}e^{-2.1} \quad = .270$$
$$P(3) = \frac{(2.1)^3}{3!}e^{-2.1} \quad = .189$$

$$P(4 \text{ or more}) = - (.122 + .257 + .270 + .189) = .162$$

Multiplying each of these probabilities by 450 gives the number of patients expected to survive each of the designated numbers of years:

$$.122(450) = 54.9, \quad .257(450) = 115.65, \quad .270(450) = 121.5,$$
$$.189(450) = 85.05, \quad .162(450) = 72.9.$$

Survival Years

	0	1	2	3	4 or more
Expected number of patients	54.9	115.65	121.5	85.05	72.9

Thus

$$\chi^2 = \frac{(60-54.9)^2}{54.9} + \frac{(110-115.65)^2}{115.65} + \frac{(125-121.5)^2}{121.5}$$
$$+ \frac{(88-85.05)^2}{85.05} + \frac{(67-72.9)^2}{72.9} = 1.430$$

H_0: good fit to a Poisson distribution with $\mu = 2.1$

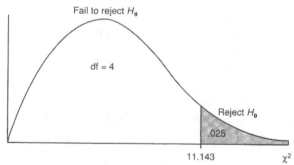

The degrees of freedom are the number of classes minus 1, that is, df = 5 − 1 = 4. With $\alpha = .025$, the critical χ^2-value is 11.143. Since 1.430 <

11.143, there is *not* sufficient evidence to reject H_0. It should *not* be disputed that a Poisson distribution with $\mu = 2.1$ describes the distribution of survival years of patients stricken with this cancer. [On the TI-83, $P = \chi^2\text{cdf}(1.430, 1000, 4) = .8390$.]

KEY EXAMPLE

A sample of 225 bags of rice labeled as containing 50 pounds each are weighed with the following results:

Weight (pounds)

	under 49.25	49.25–49.75	49.75–50.25	50.25–50.75	over 50.75
Observed number of bags	25	61	70	59	10

Test the null hypothesis that the distribution follows a normal distribution with $\mu = 50$ and $\sigma = 0.5$. Assume a 2.5% and then a 1% significance level.

Answer: The z-scores of 50.25 and 50.75 are, respectively,

$$\frac{50.25 - 50}{0.5} = 0.5 \quad \text{and} \quad \frac{50.75 - 50}{0.5} = 1.5$$

Similarly, 49.75 and 49.25 have z-scores of -0.5 and -1.5. Using Table A for normal curves we find:

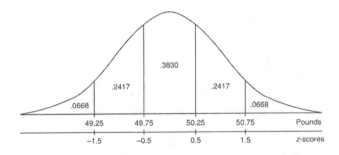

Multiplying each probability by 225 bags gives the expected numbers of bags for the corresponding weight ranges:

$$.0668(225) = 15.03, \quad .2417(225) = 54.38, \quad .3830(225) = 86.18$$

Weight (pounds)				
under 49.25	49.25–49.75	49.75–50.25	50.25–50.75	over 50.75
Expected number of bags				
15.03	54.38	86.18	34.38	15.03

Thus

$$\chi^2 = \frac{(25-15.03)^2}{15.03} + \frac{(61-54.38)^2}{54.38} + \frac{(70-86.18)^2}{86.18}$$
$$+ \frac{(59-54.38)^2}{54.38} + \frac{(10-15.03)^2}{15.03} = 12.533$$

H_0: good fit to a normal distribution with $\mu = 50$ and $\sigma = 0.5$

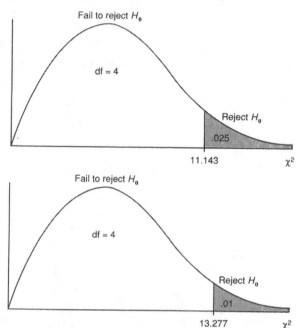

With df = 5 – 1 = 4 and $\alpha = .025$, the critical χ^2-value is 11.143, while $\alpha = .01$ results in a critical value of 13.277. Since 12.533 > 11.143 and 12.533 < 13.277, we have sufficient evidence to reject H_0 at the 2.5% significance level but not at the 1% level. Thus, if we are willing to accept a 2.5% α-risk, we should conclude that the distribution of weights of rice bags is *not* normal. [On the TI-83, $P = \chi^2\text{cdf}(12.533,1000,4) = .0138$.]

Key 63 Fits that are too good

OVERVIEW *We have been using only small, upper-tail probabilities of the chi-square distribution. There are cases, however, where we are interested in the probability that a chi-square value is less than or equal to a critical value.*

KEY EXAMPLE

Suppose your mathematics instructor gives, as a homework assignment, the problem of testing the fairness of a certain die. You are asked to roll the die 6000 times and note how often it comes up 1, 2, 3, 4, 5, and 6. The assignment tires you quickly, so you decide to simply invest some "reasonable" data. Being careful to make the total 6000, you write:

Number of times observed	1	2	3	4	5	6
	988	991	1010	990	1013	1008

H_0: good fit with uniform distribution, that is, the die if fair

The "expected" values are all 1000, and so

$$\chi^2 = \frac{12^2}{1000} + \frac{9^2}{1000} + \frac{10^2}{1000} + \frac{10^2}{1000} + \frac{13^2}{1000} + \frac{8^2}{1000} = 0.658$$

What is the conclusion if $\alpha = .01$? $.05$? $.10$?

Answer: With df $= 6 - 1 = 5$, the critical χ^2-values are 15.086, 11.070, and 9.236. Since 0.658 is less than each of these, there is not enough evidence to claim that the die is unfair.

What if we're willing to accept $\alpha = .90$? $.95$? $.975$? $.99$?

Answer: These give critical chi-square values of 1.610, 1.145, 0.831, and 0.554, so only if we accept a 99% chance of committing a Type I error can we reject H_0 and question the fairness of the die. Even if the die is fair, the probability of obtaining such a "good-fitting" sample is extremely small. Your instructor would reasonably conclude that you never completed the assignment, but simply made up the data. Your results are too good to be true! [On the TI-83, $P = \chi^2\text{cdf}(0.658, 1000, 5) = .9852$.]

Key 64 Independence and contingency
tables

OVERVIEW *In each of the goodness-of-fit problems of Keys 60–63, there was a set of expectations based on some assumption about how the distribution should turn out. We then tested whether or not an observed sample distribution might reasonably have come from a larger set based on the assumed distribution. In many real-world problems, however, we want simply to compare two or more observed samples without any prior assumptions about an expected distribution. In what is called a **test of independence**, we ask whether the two or more samples might have reasonably come from some one larger set. For example, do students, professors, and administrators all have the same opinion concerning the need for a new science building? Do non-smokers, light smokers, and heavy smokers all have the same likelihood of eventually receiving a diagnosis of cancer, heart disease, or emphysema?*

We classify our observations in two ways, and then ask whether the two ways are independent. For example, we might consider several age groups, and within each age group ask how many employees show various levels of job satisfaction. The null hypothesis would be that age and job satisfaction are independent, that is, that the proportion of employees expressing a given level of job satisfaction is the same no matter which age group is considered. A sociologist might classify people by ethnic origin, and within each group ask how many individuals complete various levels of education. The null hypothesis would be that ethnic origin and education level are independent, that is, that the proportion of people achieving a given education level is the same no matter which group is considered.

Our analysis will involve calculating a table of *expected* values, assuming the null hypothesis about independence is true. We compare these expected values with the observed values, and ask whether the differences are reasonable if H_0 is true. The significance of the differences is gauged by the same χ^2-value of weighted squared differences used in the preceding keys. The smaller the resulting χ^2-value, the more reasonable is the null hypothesis of independence. If the χ^2-value exceeds some critical value, then we can say that there is sufficient evidence to reject the null hypothesis and claim that there *is* some relationship between the two variables or methods of classification.

KEY EXAMPLE

A beef distributor wishes to determine whether there is a relationship between geographic region and cut of meat preferred. If there is no relationship, we will say that beef preference is *independent* of geographic region. Suppose that, in a random sample of 500 consumers, 300 are from the North while 200 are from the South. Of these, 150 prefer cut A, 275 prefer cut B, and only 75 prefer cut C.

| | Geographic Region | | |
	North	South	
Cut A			150
Cut B			275
Cut C			75
	300	200	

Beef Preference

If beef preference is independent of geographic region, how would we expect this table to be filled in?

Answer: Since 300/500 = .6 of the sample is from the North, we would expect .6 of the 150 consumers favoring cut A to be from the North. We calculate .6(150) = 90 or (300)(150)/500 = 90. Similarly, we would expect .6 of the 275 consumers favoring cut B to be from the North. We calculate .6(275) = 165 or (300)(275)/500 = 165. Continuing in this manner we fill in the table as follows:

Expected results:

	North	South	
Cut A	90	60	150
Cut B	165	110	275
Cut C	45	30	75
	300	200	

Note that we didn't have to actually perform all the calculations to fill in the table. After determining 90 and 165, there were really no choices left for the remaining values because the row totals and column totals were already set. Thus, in this problem there are *two* degrees of freedom, that is, df = 2.

Now suppose that in the actual sample of 500 consumers the observed numbers were as follow:

Observed results:

	North	South	
Cut A	100	50	150
Cut B	150	125	275
Cut C	50	25	75
	300	200	

Are the differences between the expected and observed values large or small? If the differences are large enough, we will reject independence and claim that beef preference is related to geographic location. We calculate the χ^2-value:

$$\chi^2 = \Sigma \frac{(obs. - exp.)^2}{exp.}$$

$$= \frac{(100 - 90)^2}{90} + \frac{(50 - 60)^2}{60} + \frac{(150 - 165)^2}{165} + \frac{(125 - 110)^2}{110}$$

$$+ \frac{(50 - 45)^2}{45} + \frac{(25 - 30)^2}{30}$$

$$= 7.578$$

Is 7.578 large enough for us to reject the null hypothesis of independence?

Answer:

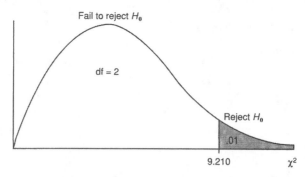

Looking at the χ^2 table, with df = 2, we see that, with α = .01, the critical χ^2 is 9.210. Thus there is not enough evidence to reject H_0.

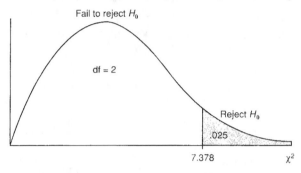

However, with α = .025, the critical χ^2 is 7.378; and since 7.578 > 7.378, there is sufficient evidence to reject H_0. Thus, we can conclude that at the 2.5% significance level there is a relationship between beef preference and geographic region, while at the 1% significance level there is *not* sufficient evidence to reject the null hypothesis of independence. [On the TI-83, go to MATRIX and EDIT. Put the observed data into a matrix. Then STAT, TESTS, χ^2-Test, will give χ^2 = 7.576 and P = .0226.]

KEY EXAMPLE

In a nationwide telephone poll of 1000 adults, representing Democrats, Republicans, and Independents, respondents were asked if their confidence in the U.S. banking system had been shaken by the Savings and Loan crisis. The answers, cross-classified by party affiliation, are given in the following *contingency* table.

Observed results:

	Yes	No	No opinion
Democrats	175	220	55
Republicans	150	165	35
Independents	75	105	20

Test the null hypothesis that shaken confidence in the banking system is independent of party affiliation. Use a 10% significance level.

Answer: The above table gives the *observed* results. To find the *expected* values, we must first determine the row and column totals, which were given in the preceding example.

We obtain the following values:

Row totals: 175 + 220 + 55 = 450, 150 + 165 + 35 = 350,
                75 + 105 + 20 = 200

Column totals: 175 + 150 + 75 = 400, 220 + 165 + 105 = 490,
                55 + 35 + 20 = 110.

	Yes	No	No opinion	
Democrats				450
Republicans				350
Independents				200
	400	490	110	

To calculate, for example, the expected value in the upper left box, we can proceed in any of several equivalent ways. First, we could note that the proportion of Democrats is 450/1000 = .45; and so, if independent, the expected number of Democrat "yes" responses is .45(400) = 180. Instead, we could note that the proportion of "yes" response is 400/1000 = .4; and so, if independent, the expected number of Democrat "yes" response is .4(450) = 180. Finally, we could note that both these calculations simply involve (450)(400)/1000 = 180. In other words, the expected value of any box can be calculated by multiplying its corresponding row total times the appropriate column total and then dividing by the grand total. Thus, for example, the expected value for the middle box corresponding to Republican "no" responses is (350)(490)/1000 = 171.5. Continuing in this manner, we find:

Expected results:

	Yes	No	No opinion	
Democrats	180	220.5	49.5	450
Republicans	140	171.5	38.5	350
Independents	80	98	22	200
	400	490	110	

Then

$$\chi^2 = \frac{(175-180)^2}{180} + \frac{(220-220.5)^2}{220.5} + \frac{(55-49.5)^2}{49.5} + \frac{(150-140)^2}{140}$$

$$+ \frac{(165-171.5)^2}{171.5} + \frac{(35-38.5)^2}{38.5} + \frac{(75-80)^2}{80} + \frac{(105-98)^2}{98}$$

$$+ \frac{(20-22)^2}{22}$$

$$= 3.024$$

Note that, when the 180, 220.5, 140, and 171.5 boxes were calculated, the other expected values could be found by using the row and column totals. Thus, the number of degrees of freedom here is 4. More generally, in this type of problem

$$df = (r-1)(c-1)$$

where r is the number of rows and c is the number of columns.

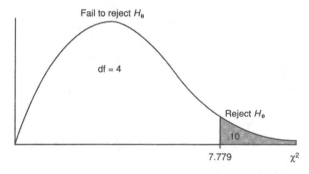

With $\alpha = .10$ and $df = 4$, the critical χ^2-value is 7.779. Since $3.024 < 7.779$, there is *not* sufficient evidence to reject the null hypothesis of independence. Thus, at the 10% significance level, we *cannot* claim a relationship between party affiliation and shaken confidence in the banking system. [On the TI-83, χ^2-Test gives $\chi^2 = 3.0243$ and $P = .5538$.]

Two points are worth noting:

- If any of the *expected* values are too small, the χ^2-value tends to turn out to be unfairly high. The usual rule of thumb cut-off point is taken to be 5; that is, this procedure is not used if any expected value is below 5. Sometimes this difficulty is overcome by a regrouping that combines two or more classifications. However, this also leads to smaller numbers of degrees of freedom and less information from the tables.

- Even when we have sufficient evidence to reject the null hypothesis of independence, we cannot necessarily claim any direct *causal* relationship. In other words, while we do make a statement about some link or relationship between two variables, we are not justified in claiming that one is causing the other. For example, we may show a relationship between salary level and job satisfaction, but our methods would not show that higher salaries *cause* higher job satisfaction. Perhaps if an employee has higher job satisfaction, this causes his superiors to be impressed and thus leads to larger increases in pay. Or perhaps there is a third variable, such as training, education, or personality, that has a direct causal relationship with both salary level and job satisfaction.

Key 65 Chi-square test for homogeneity of proportions

OVERVIEW *Chi-square procedures can be used with a single variable to compare samples from two or more distinct populations. It is important that the samples be **simple random samples**, that they be taken **independently** of each other, that the original populations be large compared with the sample sizes, and that the expected values for all cells be at least 5. The contingency table used has a row for each sample.*

KEY EXAMPLE

Data are gathered to determine which group of school employees has the greatest proportion who are satisfied with their jobs. In independent simple random samples of 100 teachers, 60 administrators, 45 custodians, and 55 secretaries, the numbers satisfied with their jobs were found to be 82, 38, 34, and 36, respectively. Is there evidence that the proportion of employees satisfied with their jobs is different in different school system job categories?

Answer: H_0: The proportion of employees satisfied with their jobs is the same across the various school system job categories.

H_a: At least two of the job categories differ in the proportion of employees satisfied with their jobs.

The observed counts as follows:

	Satisfied	Not satisfied
Teachers	82	18
Administrators	38	22
Custodians	34	11
Secretaries	36	19

Just as we did in the previous Key, we can calculate the expected value of any cell by multiplying the corresponding row total by the appropriate column total and then dividing by the grand total. In this case, this results in the following expected counts:

	Satisfied	Not satisfied	
Teachers	73.1	26.9	100
Administrators	43.8	16.2	60
Custodians	32.9	12.1	45
Secretaries	40.2	14.8	55
	190	70	260

We note that all expected cell counts are >5, and then calculate chi-square:

$$\chi^2 = \Sigma \frac{(obs - exp)^2}{exp} = \frac{(82 - 73.1)^2}{73.1} + \cdots + \frac{(19 - 14.8)^2}{14.8} = 8.640$$

With $4 - 1 = 3$ degrees of freedom, we see from Table C that 8.640 falls between tail probabilities of .025 and .05. [With the TI-83, we can calculate $P = \chi^2\text{cdf}(8.640, 1000, 4) = .0345$.] With this small a P-value, there is sufficient evidence to reject H_0, and we can conclude that there is evidence that the proportion of employees satisfied with their jobs is *not* the same across all the school system job categories. *Note:* On the TI-83 we could also have put the observed data into a matrix and used χ^2-Test, resulting in $\chi^2 = 8.707$ and $P = .0335$.

Key 66 Theme exercises with answers

OVERVIEW *Sample questions of the type that might appear on homework assignments and tests are presented with answers.*

- A pet food manufacturer runs an experiment to determine whether three brands of dog food are equally preferred. In the experiment, 150 dogs are individually set loose in front of three dishes of food and their choices are noted. Tabulations show that 62 dogs went to brand A, 43 to brand B, and 45 to brand C. Is there sufficient evidence to say that dogs have preferences among the brands? Test at the 2.5% significance level.

  Answer: If dogs have no preferences among the three brands, we would expect $150/3 = 50$ dogs to go to each dish. Thus:

  $$\chi^2 = \frac{(62-50)^2}{50} + \frac{(43-50)^2}{50} + \frac{(45-50)^2}{50} = 4.36$$

  With $3 - 1 = 2$ degrees of freedom and $\alpha = .025$, the critical χ^2-value is 7.378. Since $4.36 < 7.378$, there is *not* sufficient evidence to say that dogs have preferences among the three given brands. [Of course the company manufacturing brand A will continue to claim that more dogs prefer its food!]

- A highway superintendent states that five bridges into a city are used in the ratio 2:3:3:4:6 during the morning rush hour. A highway study of a sample of 9000 cars indicates that 1070, 1570, 1513, 1980, and 2867 cars, respectively, use the five bridges. Can the superintendent's claim be rejected if $\alpha = .05$? $.025$?

  Answer: Since $2 + 3 + 3 + 4 + 6 = 18$, according to the superintendent the expected number of cars using each bridge is as follows:

  $$\frac{2}{18}(9000) = 1000, \quad \frac{3}{18}(9000) = 1500, \quad 1500,$$
  $$\frac{4}{18}(9000) = 2000, \quad \frac{6}{18}(9000) = 3000$$

We calculate chi-square:

$$\chi^2 = \frac{(1070-1000)^2}{1000} + \frac{(1570-1500)^2}{1500} + \frac{(1513-1500)^2}{1500}$$

$$+ \frac{(1980-2000)^2}{2000} + \frac{(2867-3000)^2}{3000}$$

$$= 14.38$$

With $5 - 1 = 4$ degrees of freedom and $\alpha = .01$, the critical χ^2-value is 13.277. Since $14.38 > 13.277$, there *is* sufficient evidence to reject H_0 and to claim that the superintendent is wrong.

- Four commercial flights per day are made from a small county airport. Suppose the airport manager tabulates the number of on-time departures each day for 200 days.

Number of On-Time Departures

	0	1	2	3	4
Observed number of days	13	36	72	56	23

At the 5% significance level, test the null hypothesis that the daily distribution is binomial.

Answer: In this problem, we are not given a value of p to test against, so we use the observed sample to find a proportion $\hat{p}$. The number of on-time departures is seen to be $13(0) + 36(1) + 72(2) + 56(3) + 23(4) = 440$. Since there were $4(200) = 800$ flights, this gives a proportion of

$$\hat{p} = \frac{440}{800} = .55$$

The binomial distribution with $p = .55$ and $n = 4$ is

$$P(0) = (.45)^4 \qquad\qquad = .0410$$
$$P(1) = 4(.55)(.45)^3 \quad = .2005$$
$$P(2) = 6(.55)^2(.45)^2 = .3675$$
$$P(3) = 4(.55)^3(.45) \quad = .2995$$
$$P(4) = (.55)^4 \qquad\qquad = .0915$$

Multiplying each of these probabilities by 200 days gives the expected numbers of occurrences: .0410(200) = 8.2, .2005(200) = 40.1, .3675(200) = 73.5, .2995(200) = 59.9, .0915(200) = 18.3.

Number of On-Time Departures

	0	1	2	3	4
Expected number of days	8.2	40.1	73.5	59.9	18.3

We calculate chi-square:

$$\chi^2 = \frac{(13-8.2)^2}{8.2} + \frac{(36-40.1)^2}{40.1} + \frac{(72-73.5)^2}{73.5} + \frac{(56-59.9)^2}{59.9}$$
$$+ \frac{(23-18.3)^2}{18.3} = 4.721$$

H_0: good fit with a binomial

In this case the sample not only gives the total number, 200, but also gives the proportion, .55. Thus the number of degrees of freedom is the number of classes minus 2, that is, df = 5 − 2 = 3. With α = .05, this gives a critical χ^2-value of 7.815. Since 4.721 < 7.815, there is *not* sufficient evidence to reject H_0. The observed data do *not* differ significantly from what would be expected for a binomial distribution.

- An editor checks the number of typing errors per page on 125 pages of a manuscript.

Number of errors

	0	1	2	3	4	5
Observed number of pages	29	38	24	18	13	3

Does the number of errors follow a Poisson distribution? Test at the 10% significance level.

Answer: In this case the mean is not given, so it must be estimated from the sample. The total number of errors is 29(0) + 38(1) + 24(2) + 18(3) + 10(4) + 3(5) = 195. Thus the sample average per page is 195/125 = 1.56. The Poisson probabilities are:

$$P(0) = \quad e^{-1.56} \quad = .210$$

$$P(1) = \quad 1.56e^{-1.56} \quad = .328$$

$$P(2) = \frac{(1.56)^2}{2}e^{-1.56} \quad = .256$$

$$P(3) = \frac{(1.56)^3}{3!}e^{-1.56} \quad = .133$$

$$P(4) = \frac{(1.56)^4}{4!}e^{-1.56} \quad = .052$$

$P(5 \text{ or more}) = 1 - (.210 + .328 + .256 + .133 + .052) = .021$

Multiplying by 125 pages gives the expected number of pages with various numbers of mistakes: $.210(125) = 26.25$, $.328(125) = 41$, $.256(125) = 32$, $.133(125) = 16.625$, $.052(125) = 6.5$, and $.021(125) = 2.625$.*

Number of errors

	0	1	2	3	4	5
Expected number of pages	26.25	41	32	16.625	6.5	2.625

Thus

$$\chi^2 = \frac{(29-26.25)^2}{26.25} + \frac{(38-41)^2}{41} + \frac{(24-32)^2}{32} + \frac{(18-16.625)^2}{16.625}$$

$$+ \frac{(13-6.5)^2}{6.5} + \frac{(3-2.625)^2}{2.625}$$

$$= 9.175$$

H_0: good fit with a Poisson distribution

Since the sample is giving both the total number of pages (125) and the average errors per page (1.56), the number of degrees of freedom is the number of classes minus 2, that is, df = 6 − 2 = 4. With $\alpha = .10$, the critical χ^2-value is 7.779. Since 9.175 > 7.779, there *is* sufficient evidence at the 5% level to reject H_0 and to claim that the typing errors per page do *not* follow the Poisson distribution.

---

*Note that this value violates the previously mentioned rule of thumb that every cell must have an expected value of at least 5. Some statisticians do accept one lower value; alternatively, we could combine the last two cells into one with a higher expected value.

- Suppose that the assembly times for a sample of 300 units of an electronic product have mean $\mu = 84$, standard deviation $\sigma = 3$, and the following distribution:

Assembly Time (minutes)

	< 78	78–81	81–84	84–87	87–90	>90
Observed number of units	15	39	87	96	48	15

At the 1% significance level, test the null hypothesis that the distribution is normal.

Answer: Here we must use the sample mean, 84, and the sample standard deviation, 3. The normal probability table gives:

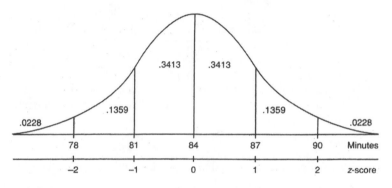

Multiplying by 300 units yields the following expected numbers of units.

Assembly Time (minutes)

	< 78	78–81	81–84	84–87	87–90	>90
Expected number of units	6.84	40.77	102.39	102.39	40.7	6.84

Thus

$$\chi^2 = \frac{(15 - 6.84)^2}{6.84} + \frac{(39 - 40.77)^2}{40.77} + \frac{(87 - 102.39)^2}{102.39} + \frac{(96 - 102.39)^2}{102.39}$$
$$+ \frac{(48 - 40.77)^2}{40.77} + \frac{(15 - 6.84)^2}{6.84}$$
$$= 23.540$$

Since we are using three measures from the sample (size, mean, and standard deviation), the number of degrees of freedom equals the number of classes minus 3, that is, df = 6 − 3 = 3. With $\alpha = .01$, the critical χ^2-value is 11.34. Since 23.540 > 11.34, there *is* sufficient evidence to reject H_0 and to conclude that the distribution of assembly times is *not* normal.

- A medical researcher tests 640 heart-attack victims for the presence of a certain antibody in their blood and cross-classifies against the severity of the attack. The results are reported in the following table:

Observed results:

Severity of Attack

		Severe	Medium	Mild
Antibody Test	Positive	85	125	150
	Negative	40	95	145

Is there evidence of a relationship between presence of the antibody and severity of the heart attack? Test at the 5% significance level.

Answer:

H_0: independence (no relation between antibody and attack)

H_a: dependence (attack severity related to presence of antibody)

$$\alpha = .05, \quad df = (2-1)(3-1) = 2$$

Totaling rows and columns yields:

	Severe	Medium	Mild	
Positive				360
Negative				280
	125	220	295	

The expected value for each box is calculated by multiplying row total by column total and dividing by 640:

Expected results:

	Severe	Medium	Mild	
Positive	70.3	123.8	165.9	360
Negative	54.7	96.2	129.1	280
	125	220	295	

Then

$$\chi^2 = \frac{(85-70.3)^2}{70.3} + \frac{(125-123.8)^2}{123.8} + \frac{(150-165.9)^2}{165.9}$$
$$+ \frac{(40-54.7)^2}{54.7} + \frac{(95-96.2)^2}{96.2} + \frac{(145-129.1)^2}{129.1}$$
$$= 10.533$$

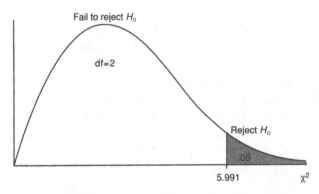

With $\alpha = .05$ and df = 2, the critical χ^2-value is 5.991. Since $10.533 > 5.991$, there is sufficient evidence to reject the null hypothesis of independence. Thus, at the 5% significance level, there *is* a relationship between presence of the antibody and severity of the heart attack. [On the TI-83, χ^2-Test gives $\chi^2 = 10.540$ and $P = .0051$.]

Theme 9   REGRESSION ANALYSIS

*M*any decisions are based on a perceived relationship between two variables. For example, a company's market share may vary directly with advertising expenditures. A person's blood pressure may increase or decrease inversely as less or more hypertension medication is taken. A student's grades will probably go up and down according to the number of hours of studying time per week.

Two questions arise. First, how can the strength of an apparent relationship be measured? Second, how can an observed relationship be put into functional terms? For example, not only may a real estate broker wish to determine whether a relationship exists between the prime rate and the number of new homes sold in a month, but also it is useful to develop an expression with which to predict the number of house sales given a particular value of the prime rate.

Key 67 Scatterplots

OVERVIEW *Suppose a relationship is perceived between two variables called X and Y, and we graph the pairs (x,y). The result, called a scatterplot gives a visual impression of the existing relationship between the variables.*

KEY EXAMPLE

The following table gives the ages and salaries (in $1000's) of four executives in a business firm.

Age	38	53	42	47
Salary	45	86	58	61

Plotting the four points (38,45), (53,86), (42,58), and (47,61) gives the following scatterplot:

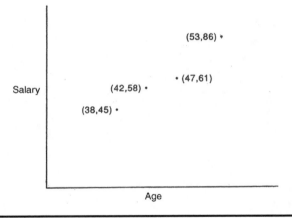

In the following keys, we will be interested in finding the *best-fitting* straight line that can be drawn through such a scatterplot.

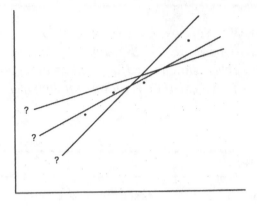

Key 68 Equation of the regression line

OVERVIEW *The best-fitting straight line, that is, the line that minimizes the sum of the squares of the differences between the observed values and the values predicted by the line, is called the* **regression line**, *(or* **least squares regression line**).

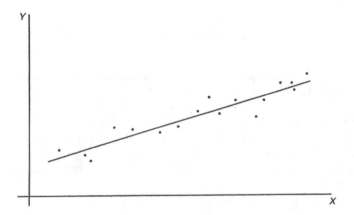

It is reasonable, intuitive, and correct that the best-fitting line will pass through $(\bar{x}, \bar{y})$, where $\bar{x}$ and $\bar{y}$ are the means of the variables X and Y. Then, from the basic expression for a line with a given slope through a given point, comes the equation

$$\hat{y} - \bar{y} = m(x - \bar{x})$$

or

$$\hat{y} = \bar{y} + m(x - \bar{x})$$

where m is the slope of the line.

It can be shown algebraically that

$$m = \frac{\sum xy - n\bar{x}\bar{y}}{\sum x^2 - n(\bar{x})^2}$$

where

$$\sum xy = x_1 y_1 + x_2 y_2 + \cdots + x_n y_n \quad \text{and} \quad \sum x^2 = x_1^2 + x_2^2 + \cdots + x_n^2$$

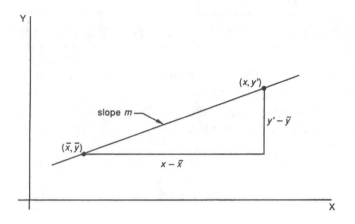

KEY EXAMPLE

An insurance company conducts a survey of 15 of its life insurance agents. The average number of minutes spent with each potential customer and the number of policies sold in a week are noted for each agent. Letting X and Y represent the average number of minutes and the number of sales, respectively, we have:

X	25	23	30	25	20	33	18	21	22	30	26	26	27	29	20
Y	10	11	14	12	8	18	9	10	10	15	11	15	12	14	11

Find the equation of the best-fitting straight line for this data.

Answer: Plotting the 15 points, (25,10), (23,11),..., (20,11), gives an intuitive visual impression of the relationship:

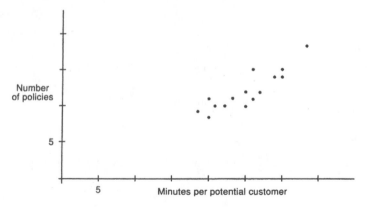

The above scatterplot indicates the existence of a relationship that appears to be *linear*; that is, the points lie roughly on a straight line. Furthermore, the linear relationship is *positive*; that is, as one variable increases, so does the other (the straight line slopes upward).

To determine the equation of the best-fitting line, we first calculate:

$$\Sigma x = 25 + 23 + 30 + \cdots + 20 = 375$$
$$\Sigma y = 10 + 11 + 14 + \cdots + 11 = 180$$
$$\Sigma x^2 = 25^2 + 23^2 + 30^2 + \cdots + 20^2 = 9639$$
$$\Sigma xy = 25(10) + 23(11) + 30(14) + \cdots + 20(11) = 4645$$

Then $n = 15$, so

$$\bar{x} = \frac{\Sigma x}{n} = \frac{375}{15} = 25, \quad \bar{y} = \frac{\Sigma y}{n} = \frac{180}{15} = 12$$

and

$$m = \frac{\Sigma xy - n\bar{x}\,\bar{y}}{\Sigma x^2 - n(\bar{x})^2} = \frac{4645 - 15(25)(12)}{9639 - 15(25)^2} = \frac{145}{264} = 0.5492$$

Thus the regression line is given by

$$\hat{y} = \bar{y} + m(x - \bar{x}) = 12 + 0.5492(x - 25) = 0.5492x - 1.73$$

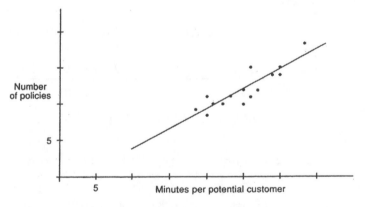

On the TI-83, put the data into L1 and L2, then STAT → CALC → LinReg($ax+b$) quickly gives $\hat{y} = 0.5492x - 1.731$.

Key 69 Slope of and predictions from the regression line

OVERVIEW *The regression line enables us to predict* Y *values from given* X *values. The slope of the regression line yields a numerical insight into the relationship between the variables.*

KEY EXAMPLE

In the example of Key 68, the regression line was calculated to be $\hat{y} = 0.5492x - 1.73$. We might predict that agents who average 24 minutes per customer will average $0.5492(24) - 1.73 = 11.45$ sales per week. We also note that each additional minute spent seems to produce an average 0.5492 number of extra sales.

KEY EXAMPLE

Following are advertising expenditures and total sales with regard to six detergent products.

Advertising ($1000's):	x	2.3	5.7	4.8	7.3	5.9	6.2
Total sales ($1000's):	y	77	105	96	118	102	95

Find the equation of the regression line. Interpret the slope.

Answer: We calculate

$$\Sigma x = 32.2, \quad \Sigma y = 593, \quad \Sigma xy = 3288.6, \quad \Sigma x^2 = 187.36$$

which give

$$\bar{x} = 5.367, \quad \bar{y} = 98.833, \quad m = 7.293$$

Thus

$$\hat{y} = 98.833 + 7.293(x - 5.367) = 7.293x + 56.691$$

Using the equation of the regression line, we can predict, for example, that if $5000 is spent on advertising, the resulting total sales will be $7.293(5) + 56.691 = 93.156$ thousand or $93,156.

The slope of the regression line indicates that on average every extra $1000 spent on advertising will result in $7293 in added sales.

Key 70 Correlation coefficient

OVERVIEW *There are ways to gauge whether or not the relationship between variables is strong enough so that finding the regression line and making use of it are meaningful.*

One measure of an apparent relationship is called the *correlation coefficient* and is denoted as r. The value r^2 is actually the ratio of the variance of the predicted values, $\hat{y}$, to the variance of the observed values, y.

It can be shown algebraically that

$$r^2 = \frac{(\sum xy - n\bar{x}\,\bar{y})^2}{\left[\sum x^2 - n(\bar{x})^2\right]\left[\sum y^2 - n(\bar{y})^2\right]}$$

There is also a relationship between r and the slope m given by $m = r\dfrac{s_y}{s_x}$ where s_y and s_x are the standard deviations of the two sets.

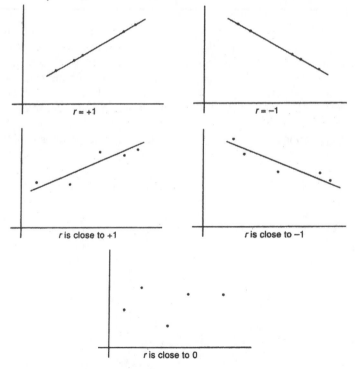

The correlation coefficient r will take the same sign as m (the slope of the regression line). The value of r always falls between -1 and $+1$, with -1 indicating perfect negative correlation, and $+1$ indicating perfect positive correlation.

KEY EXAMPLE

For the example in Key 68, we noted that $n = 15$, $\Sigma x = 375$, $\bar{x} = 25$, $\Sigma y = 180$, $\bar{y} = 12$, $\Sigma x^2 = 9639$, and $\Sigma xy = 4645$. In addition, we now calculate

$$\Sigma y^2 = 10^2 + 11^2 + 14^2 + \cdots + 11^2 = 2262$$

and thus

$$r^2 = \frac{(4645 - 15(5)(12))^2}{[9639 - 15(25)^2][2262 - 15(12)^2]} = \frac{(145)^2}{(264)(102)} = .7808$$

and $r = .8836$.

[On the TI-83, LinReg will also give r and r^2 if DiagnosticOn under Catalog is enabled.]

r^2 is called the *coefficient of determination*, is usually given as a percentage, and represents the percentage of variation in y that is explained by the variation in x. How large a value of r^2 is desirable depends on the application under consideration. Although scientific experiments often aim for an r^2 in the 90% or above range, observational studies with r^2 of 10% to 20% might be considered informative. Note that although a correlation of .6 is twice a correlation of .3, the corresponding r^2 of 36% is *four* times the corresponding r^2 of 9%.

Key 71 Residuals

OVERVIEW *The difference between an observed and predicted value is called the **residual**. When the regression line is graphed on the scatterplot, the residual of a point is the vertical distance the point is from the regression line.*

KEY EXAMPLE

We calculate the predicted values from the regression line in the Advertising/Sales Key Example in Key 69, and subtract from the observed values to obtain the residuals:

x	2.3	5.7	4.8	7.3	5.9	6.2
y	77	105	96	118	102	95
$\hat{y}$	76.5	101.3	94.7	112.9	102.7	104.9
$y - \hat{y}$	0.5	3.7	1.3	5.1	-0.7	-9.9

Note that the sum of the residuals is: $0.5 + 3.7 + 1.3 + 5.1 - 0.7 - 9.9 = 0.0$

The above equation is true in general; that is, *the sum and thus the mean of the residuals is always zero.*

Points that fall outside the overall pattern of the other points are called *outliers*. Given how residuals, $y - \hat{y}$, are defined, it is clear that outliers in the *y*-direction have large residuals. However, other outliers may or may not have large residuals.

Plotting the residuals gives further information. In particular, a residual plot with a definite pattern is an indication that a nonlinear model will show a better fit to the data than the straight regression line.

In addition to whether or not the residuals are randomly distributed, one should look at the balance between positive and negative residuals and also the size of the residuals in comparison with the associated *y*-values.

It is also important to understand that a linear model may be appropriate, but weak, with a low correlation. And, alternatively, a linear model may not be the best model (as evidenced by the residual plot), but it still might be a very good model with high r^2.

Scores whose removal would sharply change the regression line are called *influential scores*. Sometimes the description is restricted to points with extreme x-values. An influential score may have a small residual but still have a greater effect on the regression line than scores with possibly larger residuals but average x-values.

Key 72 Hypothesis test for correlation

OVERVIEW *To gauge the significance of the correlation coefficient, we may conduct a hypothesis test involving appropriate levels of significance.*

Is $r = .9636$ with four data pairs significant? How about $.8836$ with 15 data pairs? If there really is no correlation, what are the chances that the r-values could have been this large? These answers depend on what level of significance we are working with. In the Appendix, Table D gives critical r-values for both the 5% and 1% levels of significance; that is, with no correlation, there are $.05$ and $.01$ probabilities of obtaining the given critical levels. One also talks about α-risks of $.05$ and $.01$, respectively. Note that one must know the number of degrees of freedom, which in this situation is $n - 2$.

We have

$$H_0: \text{no correlation}$$

$$H_a: \text{correlation}$$

If our calculated r, in absolute value, is greater than the critical r, then we reject H_0 and say that at the given significance level there is sufficient evidence that r, the correlation coefficient, is significant.

KEY EXAMPLE

For the example carried through Keys 68–70, df $= 15 - 2 = 13$, which gives critical values of $.514$ and $.641$ at the 5% and 1% significance levels, respectively. Both $.8836 > .514$ and $.8836 > .641$, so at either of these significance levels we conclude there *is* sufficient evidence to indicate correlation.

One can also perform a hypothesis test for the value of the slope of the least squares regression line. Typically the null hypothesis will be H_0: $m = 0$, that is, that there is no linear relationship between the two variables. On the TI-83, LinRegTTest gives both the t-score and P-value for this hypothesis test.

Key 73 Theme exercises with answers

OVERVIEW *Sample questions of the type that might appear on homework assignments and tests are presented with answers.*

Following are the lengths and grades of ten research papers written for a sociology professor's class.

Length (pages):	*x*	25	32	20	28	15	34	29	30	45	35
Grade:	*y*	69	81	72	75	64	89	84	73	92	86

- Find the equation of the regression line.

- Draw a scatterplot and graph the regression line.

- Use the equation to predict the grade for a student who turns in a paper 40 pages long.

- What is the slope of the regression line, and what does it signify?

- Test for correlation at both the 5% and 1% levels of significance.

Answer: We calculate as follows:

$$\Sigma x = 25 + 32 + \cdots + 35 = 293$$

$$\Sigma y = 69 + 81 + \cdots + 86 = 785$$

$$\Sigma xy = 25(69) + 32(81) + \cdots + 35(86) = 23{,}619$$

$$\Sigma x^2 = 25^2 + 32^2 + \cdots + 35^2 = 9205$$

$$\Sigma y^2 = 69^2 + 81^2 + \cdots + 86^2 = 62{,}393$$

$$\bar{x} = \frac{293}{10} = 29.3$$

$$\bar{y} = \frac{785}{10} = 78.5$$

$$m = \frac{23{,}619 - 10(29.3)(78.5)}{9205 - 10(29.3)^2} = \frac{618.5}{620.1} = .0997$$

$$\hat{y} = 78.5 + 0.997(x - 29.3) = 0.997x + 49.3$$

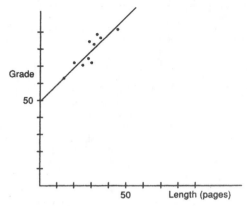

The grade for a student who turns in a paper 40 pages long is predicted to be 0.997(40) + 49.3 = 89.2.

The slope of the regression line indicates that, for each additional page, on average students can increase their grades by 0.997, that is, approximately a point per page.

$$r^2 = \frac{(618.5)^2}{(620.1)(62,393 - 10(78.5)^2)} = .801 \quad \text{and} \quad r = .895$$

The df = 10 − 2 = 8, which gives critical values of .632 and .765 at the 5% and 1% significance levels, respectively. Both .895 > .632 and .895 > .765, so at either of these significance levels we conclude that there *is* sufficient evidence to indicate correlation between a student's grade and the number of pages in the research paper.

We can also test H_0: $m = 0$ (no linear relationship) against H_a: $m \neq 0$ using LinRegTTest on the TI-83, which gives $t = 5.668$ and $P = .000471$. With such a small P-value, there is very strong evidence of a linear relationship between grade and paper length.

APPENDIX

Table A

Normal Curve Areas

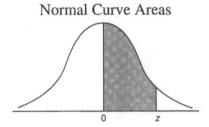

z	.00	.01	.02	.03	.04	.05	.06	.07	.08	.09
0.0	.0000	.0040	.0080	.0120	.0160	.0199	.0239	.0279	.0319	.0359
0.1	.0398	.0438	.0478	.0517	.0557	.0596	.0636	.0675	.0714	.0753
0.2	.0793	.0832	.0871	.0910	.0948	.0987	.1026	.1064	.1103	.1141
0.3	.1179	.1217	.1255	.1293	.1331	.1368	.1406	.1443	.1480	.1517
0.4	.1554	.1591	.1628	.1664	.1700	.1736	.1772	.1808	.1844	.1879
0.5	.1915	.1950	.1985	.2019	.2054	.2088	.2123	.2157	.2190	.2224
0.6	.2257	.2291	.2324	.2357	.2389	.2422	.2454	.2486	.2517	.2549
0.7	.2580	.2611	.2642	.2673	.2704	.2734	.2764	.2794	.2823	.2852
0.8	.2881	.2910	.2939	.2967	.2995	.3023	.3051	.3078	.3106	.3133
0.9	.3159	.3186	.3212	.3238	.3264	.3289	.3315	.3340	.3365	.3389
1.0	.3413	.3438	.3461	.3485	.3508	.3531	.3554	.3577	.3599	.3621
1.1	.3643	.3665	.3686	.3708	.3729	.3749	.3770	.3790	.3810	.3830
1.2	.3849	.3869	.3888	.3907	.3925	.3944	.3962	.3980	.3997	.4015
1.3	.4032	.4049	.4066	.4082	.4099	.4115	.4131	.4147	.4162	.4177
1.4	.4192	.4207	.4222	.4236	.4251	.4265	.4279	.4292	.4306	.4319
1.5	.4332	.4345	.4357	.4370	.4382	.4394	.4406	.4418	.4429	.4441
1.6	.4452	.4463	.4474	.4484	.4495	.4505	.4515	.4525	.4535	.4545
1.7	.4554	.4564	.4573	.4582	.4591	.4599	.4608	.4616	.4625	.4633
1.8	.4641	.4649	.4656	.4664	.4671	.4678	.4686	.4693	.4699	.4706
1.9	.4713	.4719	.4726	.4732	.4738	.4744	.4750	.4756	.4761	.4767
2.0	.4772	.4778	.4783	.4788	.4793	.4798	.4803	.4808	.4812	.4817
2.1	.4821	.4826	.4830	.4834	.4838	.4842	.4846	.4850	.4854	.4857
2.2	.4861	.4864	.4868	.4871	.4875	.4878	.4881	.4884	.4887	.4890
2.3	.4893	.4896	.4898	.4901	.4904	.4906	.4909	.4911	.4913	.4916
2.4	.4918	.4920	.4922	.4925	.4927	.4929	.4931	.4932	.4934	.4936
2.5	.4938	.4940	.4941	.4943	.4945	.4946	.4948	.4949	.4951	.4952
2.6	.4953	.4955	.4956	.4957	.4959	.4960	.4961	.4962	.4963	.4964
2.7	.4965	.4966	.4967	.4968	.4969	.4970	.4971	.4972	.4973	.4974
2.8	.4974	.4975	.4976	.4977	.4977	.4978	.4979	.4979	.4980	.4981
2.9	.4981	.4982	.4982	.4983	.4984	.4984	.4985	.4985	.4986	.4986
3.0	.4987	.4987	.4987	.4988	.4988	.4989	.4989	.4989	.4990	.4990

Abridged from Table I of A. Hald, *Statistical Tables and Formulas* (New York: John Wiley & Sons, Inc.), 1952. Reproduced by permission of A. Hald and the publisher, John Wiley & Sons, Inc.

Table B

Critical Values of t

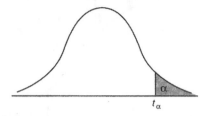

df	$t_{.100}$	$t_{.050}$	$t_{.025}$	$t_{.010}$	$t_{.005}$	$t_{.001}$	$t_{.0005}$
1	3.078	6.314	12.706	31.821	63.657	318.31	636.62
2	1.886	2.920	4.303	6.965	9.925	22.326	31.598
3	1.638	2.353	3.182	4.541	5.841	10.213	12.924
4	1.533	2.132	2.776	3.747	4.604	7.173	8.610
5	1.476	2.015	2.571	3.365	4.032	5.893	6.869
6	1.440	1.943	2.447	3.143	3.707	5.208	5.959
7	1.415	1.895	2.365	2.998	3.499	4.785	5.408
8	1.397	1.860	2.306	2.896	3.355	4.501	5.041
9	1.383	1.833	2.262	2.821	3.250	4.297	4.781
10	1.372	1.812	2.228	2.764	3.169	4.144	4.587
11	1.363	1.796	2.201	2.718	3.106	4.025	4.437
12	1.356	1.782	2.179	2.681	3.055	3.930	4.318
13	1.350	1.771	2.160	2.650	3.012	3.852	4.221
14	1.345	1.761	2.145	2.624	2.977	3.787	4.140
15	1.341	1.753	2.131	2.602	2.947	3.733	4.073
16	1.337	1.746	2.120	2.583	2.921	3.686	4.015
17	1.333	1.740	2.110	2.567	2.898	3.646	3.965
18	1.330	1.734	2.101	2.552	2.878	3.610	3.922
19	1.328	1.729	2.093	2.539	2.861	3.579	3.883
20	1.325	1.725	2.086	2.528	2.845	3.552	3.850
21	1.323	1.721	2.080	2.518	2.831	3.527	3.819
22	1.321	1.717	2.074	2.508	2.819	3.505	3.792
23	1.319	1.714	2.069	2.500	2.807	3.485	3.767
24	1.318	1.711	2.064	2.492	2.797	3.467	3.745
25	1.316	1.708	2.060	2.485	2.787	3.450	3.725
26	1.315	1.706	2.056	2.479	2.779	3.435	3.707
27	1.314	1.703	2.052	2.473	2.771	3.421	3.690
28	1.313	1.701	2.048	2.467	2.763	3.408	3.674
29	1.311	1.699	2.045	2.462	2.756	3.396	3.659
30	1.310	1.697	2.042	2.457	2.750	3.385	3.646
40	1.303	1.684	2.021	2.423	2.704	3.307	3.551
60	1.296	1.671	2.000	2.390	2.660	3.232	3.460
120	1.289	1.658	1.980	2.358	2.617	3.160	3.373
∞	1.282	1.645	1.960	2.326	2.576	3.090	3.291

This table is reproduced with the kind permission of the Trustees of Biometrika from E. S. Pearson and H. O. Hartley (eds.), *The Biometrika Tables for Statisticians*, Vol. 1, 3d ed., *Biometrika*, 1966.

Table C

The χ^2-distribution

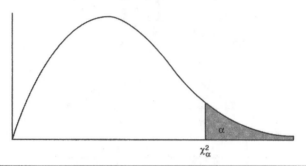

χ^2_α

df	$\chi^2_{.995}$	$\chi^2_{.990}$	$\chi^2_{.975}$	$\chi^2_{.950}$	$\chi^2_{.900}$	$\chi^2_{.100}$	$\chi^2_{.050}$	$\chi^2_{.025}$	$\chi^2_{.010}$	$\chi^2_{.005}$
1	.0000	.0002	.0010	.0039	.0158	2.706	3.841	5.024	6.635	7.879
2	.0100	.0201	.0506	.1026	.2107	4.605	5.991	7.378	9.210	10.60
3	.0717	.1148	.2158	.3518	.5844	6.251	7.815	9.348	11.34	12.84
4	.2070	.2971	.4844	.7107	1.064	7.779	9.448	11.14	13.28	14.86
5	.4117	.5543	.8312	1.145	1.610	9.236	11.07	12.83	15.09	16.75
6	.6757	.8721	1.237	1.635	2.204	10.64	12.59	14.45	16.81	18.55
7	.9893	1.239	1.690	2.167	2.833	12.02	14.07	16.01	18.48	20.28
8	1.344	1.647	2.180	2.732	3.490	13.36	15.51	17.53	20.09	21.95
9	1.735	2.088	2.700	3.325	4.168	14.68	16.92	19.02	21.67	23.59
10	2.156	2.558	3.247	3.940	4.865	15.99	18.31	20.48	23.21	25.19
11	2.603	3.053	3.816	4.575	5.578	17.27	19.68	21.92	24.72	26.76
12	3.074	3.571	4.404	5.226	6.304	18.55	21.03	23.34	26.22	28.30
13	3.565	4.107	5.009	5.892	7.042	19.81	22.36	24.74	27.69	29.82
14	4.075	4.660	5.629	6.571	7.790	21.06	23.68	26.12	29.14	31.32
15	4.601	5.229	6.262	7.261	8.547	22.31	25.00	27.49	30.58	32.80
16	5.142	5.812	6.908	7.962	9.312	23.54	26.30	28.85	32.00	34.27
17	5.697	6.408	7.564	8.672	10.09	24.77	27.59	30.19	33.41	35.72
18	6.265	7.015	8.231	9.390	10.86	25.99	28.87	31.53	34.81	37.16
19	6.844	7.633	8.907	10.12	11.65	27.20	30.14	32.85	36.19	38.58
20	7.434	8.260	9.591	10.85	12.44	28.41	31.41	34.17	37.57	40.00
21	8.034	8.897	10.28	11.59	13.24	29.62	32.67	35.48	38.93	41.40
22	8.643	9.542	10.98	12.34	14.04	30.81	33.92	36.78	40.29	42.80
23	9.260	10.20	11.69	13.09	14.85	32.01	35.17	38.08	41.64	44.18
24	9.886	10.86	12.40	13.85	15.66	33.20	36.42	39.36	42.98	45.56
25	10.52	11.52	13.12	14.61	16.47	34.38	37.65	40.65	44.31	46.93
30	13.79	14.95	16.79	18.49	20.60	40.26	43.77	46.98	50.89	53.67
40	20.71	22.16	24.43	26.51	29.05	51.81	55.76	59.34	63.69	66.77
50	27.99	29.71	32.36	34.76	37.69	63.17	67.51	71.42	76.15	79.49
60	35.53	37.48	40.48	43.19	46.46	74.40	79.08	83.30	88.38	91.95
70	43.27	45.44	48.76	51.74	55.33	85.53	90.53	95.02	100.4	104.2
80	51.17	53.54	57.15	60.39	64.28	96.58	101.9	106.6	112.3	116.3
90	59.20	61.75	65.65	69.13	73.29	107.6	113.1	118.1	124.1	128.3
100	67.33	70.66	74.22	77.93	82.86	118.5	124.3	129.6	135.8	140.2

Adapted with permission from *Biometrika Tables for Statisticians*, Vol. 1, 3d ed., Cambridge University Press, 1966, edited by E. S. Pearson and H. O. Hartley.

Table D

Critical levels of r at 5% and 1% levels of significance

df	$r_{.05}$	$r_{.01}$		df	$r_{.05}$	$r_{.01}$
1	.997	1.000		24	.388	.496
2	.950	.990		25	.381	.487
3	.878	.959		26	.374	.478
4	.811	.917		27	.367	.470
5	.754	.874		28	.361	.463
6	.707	.834		29	.355	.456
7	.666	.798		30	.349	.449
8	.632	.765		35	.325	.418
9	.602	.735		40	.304	.393
10	.576	.708		45	.288	.372
11	.553	.684		50	.273	.354
12	.532	.661		60	.250	.325
13	.514	.641		70	.232	.302
14	.497	.623		80	.217	.283
15	.482	.606		90	.205	.267
16	.468	.590		100	.195	.254
17	.456	.575		125	.174	.228
18	.444	.561		150	.159	.208
19	.433	.549		200	.138	.181
20	.423	.537		300	.113	.148
21	.413	.526		400	.098	.128
22	.404	.515		500	.088	.115
23	.396	.505		1000	.062	.081

Reproduced by the courtesy of the author and of the publisher from G. W. Snedecor and W. G. Cochran, *Statistical Methods*. The Iowa State University Press, Ames, Iowa, 1967, Table A11, p. 557.

GLOSSARY

Included here are the definitions of many, but not all, of the terms used in the keys. For formulas, calculations, and examples of these terms, and for terms not listed here, please consult the index for page references.

α-risk The probability of committing a Type I error

Bar chart A visual representation of data in which frequencies of different results are indicated by the heights of bars representing these results.

Bias A tendency to favor the selection of certain members of a population.

Binomial probabilities Probabilities resulting from applications in which a two-outcome situation is repeated some number of times, and the probability of each of the two outcomes remains the same for each repetition.

Boxplot A visual representation of dispersion that shows the largest value, the smallest value, the median, the median of the top half of the set, and the median of the bottom half of the set.

β-risk The probability of committing a Type II error

Central Limit Theorem Pick n sufficiently large (at least 30), take all samples of size n, and compute the mean of each of these samples. Then the set of these sample means will be approximately *normally* distributed.

Chebyshev's theorem For any set of data at least $(1 - 1/k^2)$ of the values lie within k standard deviations of the mean.

Chi-square A probability distribution used here in tests for goodness of fit, independence, and homogeneity of proportions.

Coefficient of determination The percentage of variation in y that is explained by the variation in x.

Confidence interval The range of values that could be taken at a given significance level.

Correlation coefficient A measure of the relationship between two variables.

Critical value A value used as a threshold to decide whether or not to reject the null hypothesis.

Empirical rule In symmetric "bell-shaped" data, about 68% of the values lie within one standard deviation of the mean, about 95% lie within two standard deviations of the mean, and more than 99% lie within three standard deviations of the mean.

Expected value (or **average** or **mean**) for a discrete random variable X, this is the sum of the products obtained by multiplying each value x by the corresponding probability $P(x)$.

Five-number summary The minimum, first quartile, median, third quartile, and maximum of a distribution.

Histogram A visual representation of data in which relative frequencies are represented by relative areas.

Mean Result of summing the values and dividing by the number of values.

Median The middle number when a set of numbers is arranged in numerical order. If there are an even number of values, the median is the result of adding the two middle values and dividing by two.

Mode The most frequent value.

Normal distribution A particular bell-shaped, symmetric curve with an infinite base.

Null hypothesis A claim to be tested, often stated in terms of a specific value for a population parameter.

Operating characteristic curve A graphical display of β values.

Outlier An extreme value falling far from most other values.

Percentile ranking Percent of all scores that fall below the value under consideration.

Poisson distribution A probability distribution that can be viewed as the limiting case of the binomial when n is large and p is small.

Population Complete set of items of interest.

Power curve The graph of probabilities that a Type II error is not committed.

Probability A mathematical statement about the likelihood of an event occurring.

P-value The probability of obtaining a sample statistic as extreme as the one obtained if the null hypothesis is assumed to be true.

Random sample When the sample is selected under conditions such that each element of the population has an equal chance to be included.

Range The difference between the largest and smallest values of a set.

Regression line The best-fitting straight line, that is, the line that minimizes the sum of the squares of the differences between the observed values and the values predicted by the line.

Residual The difference between an observed and predicted value.

Sample Part of a population used to represent the population.

Scatterplot A visual display of the relationship between two variables.

Significance level The choice of α-risk in a hypothesis test.

Simple random sample A sample selected in such a way that every element has an equal chance of being selected, and every possible sample of the desired size has an equal chance of being selected.

Simple ranking After arranging the elements in some order, noting where in the order a particular value falls.

Skewed A distribution spread thinly on one end.

Standard deviation The square root of the variance.

Stemplot A pictorial display giving the shape of the histogram, as well as indicating the values of the original data.

Student t-distribution A bell-shaped, symmetrical curve that is lower at the mean, higher at the tails, and more spread out than the normal distribution.

Type I error The error of mistakenly rejecting a true null hypothesis.

Type II error The error of mistakenly failing to reject a false null hypothesis.

Variance The average of the squared deviations from the mean.

Z-score The number of standard deviations a value is away from the mean.

INDEX